An exhaustive study of the topic… an objective presentation of conflicting opinions … a comprehensive analysis of textual, contextual and intellectual arguments… a profound contribution towards solving a long awaited international problem…an honest realization of Shariʿah objectives…

– Prof. Dr. Abdel Majid al-Najjar
Assistant Secretary General, European Council of Fatwa & Research, France

An example of honest, profound, meticulous research…a creative analysis of varying juristic opinions, a thorough demonstration of related arguments, a comprehensive comparison and preponderance of internal reasoning, a logical rather literal treatment of the subject … an erudite work deserving appreciation and esteem.

– Prof. Dr. Hussain Hamid Hassan
Former President IIU, Pakistan; Professor of Shariʿah Cairo University
CEO, Dar al-Shariʿah, Dubai

I used to waver regarding confirmation of Ramadan by astronomical calculations but no more…this book has resolved the confusion once for all…a convincing scholarly treatise… a celebrated work deserving salutation.

– Prof. Dr. Hasan al-Shaf'ai
Former President of IIU, Pakistan, Professor al-Azhar Universty
Member Academy of Arabic Language, Cairo

An enormous research work proving that the Qur'an and Sunnah never prohibited use of precise astronomical calculations, that there exists no consensus among jurists vis a vis barring use of calculations in confirming Ramadan…proving that use of calculations are in line with Islamic Shari'ah…a precisely pointing work, an abundantly rich exposition, a noble work loaded with prospects of Muslim unity especially in the West…

– Prof. Dr. Ali al-Qurrah Daghi
Professor University of Qatar
Member European Council of Fatwa & Research, Doha

This pleasantly erudite book will force many Muslims to change their views about astronomical calculations and confirmation of Ramadan. The author's studious treatment of the subject leaves no room for doubt that our precise astronomical calculations are altogether different from the guess works of medieval magicians and astrologers. Islamic Fiqh is against astrology and not against the science of astronomy.

– Prof. Dr. Mahmood Mazrooa'h
Former President, Ulama Council of Egypt,
Professor, Higher Studies, Umm al-Qura Unversity, Makkah
and al-Azhar Unversity, Cairo

The Astronomical Calculations and Ramadan: A *Fiqhī* Discourse

Zulfiqar Ali Shah

THE ASTRONOMICAL CALCULATIONS AND RAMADAN:

A Fiqhī *Discourse*

ZULFIQAR ALI SHAH

With a Preface by
MUZAMMIL H. SIDDIQI

1401AH—1981AC

THE INTERNATIONAL INSTITUTE OF ISLAMIC THOUGHT
WASHINGTON • LONDON

THE INTERNATIONAL INSTITUTE OF ISLAMIC THOUGHT
P.O. BOX 669, HERNDON, VA 22070, USA
WWW.IIIT.ORG

LONDON OFFICE
P.O. BOX 126. RICHMOND, SURREY TW9 2UD, UK
WWW.IIITUK.COM

ISBN 978-1-56564-334-5 paperback

Library of Congress Cataloging-in-Publication Data

Shah, Zulfiqar Ali.
The astronomical calculations and Ramadan: a fiqhi discourse / Zulfiqar Ali Shah ; with a preface by Muzammil H. Siddiqi.
p. cm.
Includes bibliographical references.
ISBN 978-1-56564-334-5
1. Ramadan. 2. New moon--Religious aspects--Islam. 3. New moon--Visibility. 4. Calendar, Islamic. I. Siddiqi, Muzammil. II. Title.
BP186.4.S47 2009
297.3'62--dc22
2009028286

Table of Contents

FOREWORD i

PREFACE v

INTRODUCTION 1

Chapter 1: The Cultural Milieu and Its *Fiqhī* Relevance 7
The Debate over the Prayer Timings and the Direction of the Qiblah 8
The Debate about the Unity or Variety of the Horizons 12
An Analysis of the Selective Insistence and the Issue of *Ta'wīl* 19
Allegorical Interpretations and Christianity 20
Allegorical Interpretations and Some Muslim Sects 22
Allegorical Interpretations and Muslim Philosophers 25
Allegorical Interpretations and the *Salaf* 26

Chapter 2: Analysis 30
The Qur'an Never Required Naked-Eye Sighting of the Moon 30
The Qur'anic Use of *Shahida* 31
Some Wrong Interpretations of *Shahida* 35
The *Mawāqīt* Argument 56
Moon-Sighting Supplications Are Based upon Weak Reports 57

Chapter 3: Analysis of the Hadith Arguments 61
The Sunnah in Reality Requires "Certainty": The Legal Cause Is the "Coming of Ramadan" 61
Multiple Connotations of *Ra'ā* 71
Defining the Islamic Legal Cause 80

Chapter 4: A Brief Historical Exposition of the Classical Debate about Sighting the Moon and Calculations 85
Classical Islamic Scholarship and Naked-Eye Sighting as the Cause of Ramadan 91
The Muslim Ummah Is Unlettered 94
A Summary of the Classical Majority's Arguments against Calculations 103

Chapter 5: Weakness of the Argument for *Ikmāl* (Completing Thirty) Days 106
Significance of the Variations in Transmission 114
Some Very Practical Challenges 121
Weakness of the Ijma[c] (Consensus) Argument 122

Chapter 6: Arguments for Astronomical Calculations 133
The Moon: The Divine Source of Precise Calculations 133
The Three Accepted Interpretations of These Hadiths 134
The Hadith of the Dajjal 136
The "Estimation" Interpretation of the Hadith Is More Accurate 138
Modern Jurists and Calculations 143
Conclusions 150

Endnotes 152

Bibliography 167

FOREWORD

The work before you is an important contribution to Islam's body of knowledge. It attempts to answer, quite exhaustively I might add, some of the main arguments against the Fiqh Council of North America's decision to determine holy days through computation rather than naked-eye sighting. This is a particularly onerous task, not simply because it breaks with scholarly tradition, but because tradition, being remarkably familiar with both computation's method and logic, was able to vigorously argue against it centuries ago. In those arguments, an example of which appears in the fatawa of Ibn Taymiyyah, scholars go only so far as to admit that lunar phases can indeed be computed accurately. They all hasten to add, however, that the same is not true of predicting physical sightings, which for them was the only criterion for establishing Islamic dates. As will be shown in the chapters ahead, some very notable scholars, among them Ibn Qutaybah, Ibn Surayj, and Taj al-Din al-Subki, were advocates of computation in one form or another. But for them calculations were no more than a corroboration of a naked-eye sighting, and not an independent alternative to it. The arguments that follow therefore differ in that they are meant to show the juridical validity of the computational method as well as the practical and social benefits it offers.

This brings me to computation's other benefit, that of reducing the uncertainties of religious life in the West, particularly with regard to the holy days. For Muslims in the past as well as for the large majority today that still lives as a majority, waiting for news of the moon is part of the festivities of these holy days, and not seeing it means no more than the first fast being delayed one day. But this is not problematic because it affects the nation as a whole. The social disruptions that occur as a result affect everyone equally, in the work force, at school, and on travel. The same is not true of Muslims who live in the West. In addition to all of the foregoing inconveniences they shoulder waiting for "news of the moon," there is also the added embarrassment of explaining to friends and colleagues one additional oddity of their faith. This shift to computation as the method of determining the holy days of Islam will definitely be welcomed by such people.

Two further contributions are more academic than social and apply, as such, to enhancing the overall literacy of Muslims today. The first of these is in its comprehensive citation of the classical sources; the second is in its accurate and pithy analysis of the major scholarly arguments vis-à-vis computation. The author has done readers a great service by digging deep into the Qur'an as well as the prophetic traditions to quote verbatim the relevant evidence and to then translate such evidence into English. This is a tedious task but one that is quite necessary, I might add, if only because it helps reassure the reader that no shred of relevant textual evidence has been overlooked. Any scholar or lay person wanting to know what the Qur'an and the Traditions say on moon-sighting will find easy access to both the Arabic texts as well as their translations.

As for the arguments of Muslim scholarship, these give readers unfamiliar with Muslim scholarship a meaningful glimpse into the method and rationale of the traditional scholar. It introduces them to the building blocks of Islamic law, to juridical consensus (ijmaʿ), for instance, its multiple iterations within scholarly circles, and its various usages in this and ohter discussions in the past. It includes discussions on the etymological and syntactical aspects of language, the use of words in their literal and allegorical senses, and the differences between exclamatory and declarative statements. Such discussions serve not only to educate Muslims as to the complexities of the interpretative process, but also to caution them against hastily drawing legal conclusions without the requisite training and skill.

This current discussion on computation, which touches almost all Muslims to some degree, has succeeded in sucking in opinions even from those who are otherwise disinclined to speak their minds. This is particularly true in the West, where Muslims are generally more independently minded and better educated but lacking in political authority. Those who do weigh in go all the way – from theologians perturbed by this latest sacrifice of tradition at the alter of modernity to lay Muslims who would much rather abandon that tradition than spend another sleepless night awaiting lunar confirmation.

On the other hand, for those in Karachi, Istanbul, and Damascus who live their religious lives as a majority, them such discussions are either academic and arcane, or superfluous and inconsequential to their daily grinds. In such countries it is the state, with or without the concurrence of the scholars, that determines the beginning and the end of Ramadan. Even in case where scholarship enjoys limited autonomy, or where multiple legal opinions on moon sighting are tolerated, official government announcements that endorse one opinion over all others helps remove the cloud of uncertainty found in the West.

Scholars worry however, that this effort toward greater unity through computation is, if anything, making that effort itself all the more elusive. Abandoning tradition in determining the beginning and end of Ramadan, they argue, means breaking with the juridical consensus of posterity and with the sacred teachings of the Prophet. But such breaks are far from unique, I would argue; if anything, they occured so frequently in the past that they became a defining feature of our juridical legacy. In early times, for instance, scholars differed in their interpretation and application of the *ṣūmū* (fast sighting the moon) hadith. One group, following a rule that came to be known as *ikhtilāf al-maṭāli'* (diversity of the horizons) restricted sightings to people of that specific vicinity, while another extended any single sighting to parts of the world. Today that same regional difference has undergone a further adjustment in accordance with the political changes Muslim countries have undergone since colonialism. Sightings today follow fault lines that are almost exclusively political; as such it does not matter that parts of Saudi Arabia and Yemen share the same longitudinal space and ought to be treated, scientifically speaking, as a single geographical unit. But political borders established during colonial times by the British, for example, the French, or the Dutch are what act as *de jure* cut off points beyond which sightings will not apply. There is no scriptural evidence to support such rulings, and yet even the most conservative scholars seems to find nothing untoward about this interpretation of the *sumu* hadith!

And it does not end there: rulings on moon sighting today are as unprecedented and innovative as the times in which we live. Take the ʿId after hajj ruling, for instance: never in the history of Islamic law has any scholar even remotely suggested that ʿId al-Adha' be performed on the day after ʿArafah in Arabia. And yet a sizeable group of prominent scholars is doing just that by advocating that Muslims celebrate it in accordance with the hajj ceremonies in Makkah. The ruling itself, in my opinion, is a step in the right direction – not because it tracks perfectly with some moment in our early history, but because it more adequately addresses the challenges we as a community now face in this global village. It helps unite Muslims around their high holy days at a time when the community is more diverse and widely distributed than ever.

This attempt at unity, most scholars would agree, is as much a religious responsibility as is the fulfillment of the ritual obligations themselves. All the core rituals of Islam, in fact, serve as much as symbols of each person's quest for proximity to his or her fellow human being than they do as symbols of each person's quest for proximity to God. This is why the daily prayer is almost always more meritorious when performed in congregation, why Friday congregational prayers cannot but be performed in congregation, why such prayers must be performed facing Makkah, and why they must conform (more or less) to the same physical movements.

With all of this emphasis on the symbolic value of unity, one finds inexplicable the tendency in scholarly circles to reduce unity to no more than a supererogatory ideal to be pursued only if particular rules and regulations remain inviolate. This, then, is the challenge legal scholarship faces: how best to interpret the law's secondary elements such that they do not hamper the pursuit of Islam's greater objectives?

Dr. Muneer Fareed
Former Secretary General, ISNA
Member, Fiqh Council of North America

PREFACE

THE REGULARITY AND PERMISSIBILITY OF USING THE CALCULATION METHOD IN DETERMINING THE CRESCENT IN ACCORDANCE WITH THE HIGHER OBJECTIVES OF THE SHARIʿAH

(هُوَ الَّذِي جَعَلَ الشَّمْسَ ضِيَاء وَالْقَمَرَ نُورًا وَقَدَّرَهُ مَنَازِلَ لِتَعْلَمُواْ عَدَدَ السِّنِينَ وَالْحِسَابَ مَا خَلَقَ اللّهُ ذَلِكَ إِلاَّ بِالْحَقِّ يُفَصِّلُ الآيَاتِ لِقَوْمٍ يَعْلَمُونَ.) (يونس: 5)

It is He who made the sun a shining radiance and the moon a light and measured out phases for it so that you might know the number of years and calculate time. Allah did not create all this without a true purpose. He explains His signs for those who understand. (10:5)

(الشَّمْسُ وَالْقَمَرُ بِحُسْبَانٍ.) (الرحمن: 5)

The sun and the moon follow courses (exactly) computed. (55:5)

Allah created the sun and moon, both of which move according to precise and predefined patterns. Knowledge of these patterns help people calculate time, days, and years. This pattern is not only for Ramadan, but also for all time periods throughout the year. Allah tells us that we should pray our daily salah and begin and end our fasts based on the sun's movement.

(أَقِمِ الصَّلاَةَ لِدُلُوكِ الشَّمْسِ إِلَى غَسَقِ اللَّيْلِ وَقُرْآنَ الْفَجْرِ إِنَّ قُرْآنَ الْفَجْرِ كَانَ مَشْهُودًا.) (الإسراء: 78)

Perform regular prayers in the period from the time the sun is past its zenith till the darkness of the night, and recite (the Qur'an) at dawn – dawn recitation is always witnessed. (17:78)

(....وَكُلُواْ وَاشْرَبُواْ حَتَّى يَتَبَيَّنَ لَكُمُ الْخَيْطُ الأَبْيَضُ مِنَ الْخَيْطِ الأَسْوَدِ مِنَ الْفَجْرِ ثُمَّ أَتِمُّواْ الصِّيَامَ إِلَى الَّليْلِ....) (البقرة: 187)

....and eat and drink, until the white thread of dawn appear to you distinct from its black thread; then complete your fast till the night appears.... (2:187)

Allah also told us that the crescent moons (of various months) are for people to observe time, especially the time of hajj:

(يَسْأَلُونَكَ عَنِ الْأَهِلَّةِ قُلْ هِيَ مَوَاقِيتُ لِلنَّاسِ وَالْحَجِّ.) (الْبَقَرَةِ: 189)

They ask thee concerning the new moons. Say: They are but signs to mark fixed periods of time in (the affairs of) men, and for Pilgrimage. (2:189)

The Prophet (ṢAAS)[*] explained how to observe the sun's movement to establish the times of the salah and the beginning and end of the daily fasts and of Ramadan. For centuries, Muslims observed the sun's movement with their naked eyes every day for their five daily prayers. When clocks were invented, they started to calculate its movement. They established the timings of the salah and developed perpetual salah timetables for the whole year. Now, instead of physically watching its movement, Muslims follow a timetable based on its calculated movements.

Timetables sometimes differ on the basis of the fiqh of prayer times. Some Muslims have determined that *fajr* should start when the sun is 18° below the horizon; others have determined that it should start when the sun is 15° below the horizon. Some calculate *maghrib* as beginning when the sun sets on the horizon; others calculate it as beginning when the red twilight disappears. There are similar differences with regard to the beginning and end of periods of fasting. However, Muslims have accepted the calculation of salah times and the preparation of salah timetables for the convenience of people.

Similar to this is using the crescent moon for determining the Islamic dates. The Prophet told us that we should see the crescent to begin and end the month of Ramadan:

سَمِعْتُ أَبَا هُرَيْرَةَ رَضِيَ اللَّهُ عَنْهُ يَقُولُ قَالَ النَّبِيُّ صَلَّى اللَّهُ عَلَيْهِ وَسَلَّمَ أَوْ قَالَ قَالَ أَبُو الْقَاسِمِ صَلَّى اللَّهُ عَلَيْهِ وَسَلَّمَ صُومُوا لِرُؤْيَتِهِ وَأَفْطِرُوا لِرُؤْيَتِهِ فَإِنْ غُبِّيَ عَلَيْكُمْ فَأَكْمِلُوا عِدَّةَ شَعْبَانَ ثَلَاثِينَ.[1]

Fast with sighting it [moon] and break the fast with sighting it. Complete thirty days of Shaʿban if it is cloudy.

He also said:

عَنْ نَافِعٍ عَنْ عَبْدِ اللَّهِ بْنِ عُمَرَ رَضِيَ اللَّهُ عَنْهُمَا أَنَّ رَسُولَ اللَّهِ صَلَّى اللَّهُ عَلَيْهِ وَسَلَّمَ ذَكَرَ رَمَضَانَ فَقَالَ لَا تَصُومُوا حَتَّى تَرَوُا الْهِلَالَ وَلَا تُفْطِرُوا حَتَّى تَرَوْهُ فَإِنْ غُمَّ عَلَيْكُمْ فَاقْدُرُوا لَهُ[2]

---

[*]   ṢAAS: Many Muslims say *ṣallā Allāh ʿalayhi wa sallam* (may the peace and blessings of Allah be upon him) after mentioning the Prophet.

Do not fast until you see the crescent and do not break the fast until you see it. Estimate about it in case it is cloudy.

The goal was that everyone should begin and end Ramadan with full confidence that the month has begun and that Muslims should be united in observing their fasts and celebrating their ʿId. Why did the Prophet emphasize the observance of the *hilāl*? He answered that in another authentic hadith:

حَدَّثَنَا سَعِيدُ بْنُ عَمْرٍو أَنَّهُ سَمِعَ ابْنَ عُمَرَ رَضِيَ اللَّهُ عَنْهُمَا عَنْ النَّبِيِّ صَلَّى اللَّهُ عَلَيْهِ وَسَلَّمَ أَنَّهُ قَالَ
إِنَّا أُمَّةٌ أُمِّيَّةٌ لا نَكْتُبُ وَلا نَحْسُبُ الشَّهْرُ هَكَذَا وَهَكَذَا يَعْنِي مَرَّةً تِسْعَةً وَعِشْرِينَ وَمَرَّةً ثَلاثِينَ.[3]

We are an unlettered people; we do not know how to write and how to calculate. The month is thus and thus, meaning either twenty-nine days or thirty days.

The Prophet was fully aware of his people's condition when he gave them these instructions. He did not impose a burden upon them beyond their capacity, because he wanted them to begin and end their fast and enjoy their ʿId with ease and convenience.

In general, Muslims continued to sight the crescent to determine the month of Ramadan and celebrate their ʿIds. We also have on record the statements of some early scholars and jurists who, after learning the science of astronomy, indicated that calculating the crescent can be used for Ramadan. Imam Taqi al-Din al-Subki, a great Shafiʿi jurist, even said that calculations were more reliable than naked-eye sighting.

After Muslims acquired a more advanced knowledge of the astronomical sciences, more voices were raised to rely on calculating the crescent instead of physical sighting. Most jurists did not accept calculations due to their uncertainty of whether they were correct or could be trusted. With the further development of the astronomical sciences in the last one hundred years, more and more voices are being raised by jurists in support of calculations. The famous *muḥaddith* Shaykh Ahmad Muhammad Shakir wrote a long article emphasizing that calculation is both permissible and is the most appropriate method of determining the lunar months.

The objective of the Shariʿah is that Muslims begin and end the month of Ramadan with assurance and be united in their observance of these blessed times. The objective does not seem to engage Muslims in moon sightings or to remain uncertain about the times of *ʿibādah* until the last minute. The astronomical sciences are highly advanced today, and more reliable methods are available to determine the beginning of the lunar months. On the basis of the principles of the Shariʿah, just as timetables for salah and *ṣiyām* are prepared, it is possible to prepare calendars for the lunar months and the beginning and end of Ramadan. This knowledge is now easily available and can be used.

It is ironic that instead of taking advantage of this knowledge and making things easy, the *hilāl* issue has become a very controversial and divisive matter. In Muslim countries, official bodies make the decision. Some people differ, but they have no choice except to follow the official decision. In countries where Muslim minorities live, there is more division. In Western Europe and North America, moon sighting has become far more divisive.

We thank Dr. Zulfiqar Ali Shah, the Executive Director of the Fiqh Council of North America, who has collected the basic evidence from the Qur'an and the Sunnah, as well as the linguistic usages and juristic discussions on this subject. He has given us the arguments of both sides: those who emphasize physical sighting and those who allow calculation. It is clear that our Shariʿah is flexible in this matter, and we are allowed to adopt a method that meets the rules' objectives and makes things easy for Muslims.

عن أبي هريرة عن النبي صلى الله عليه وسلم قال: إن الدين يسر ولن يشاد الدين أحد إلا غلبه فسددوا وقاربوا وأبشروا واستعينوا بالغدوة والروحة وشيء من الدلجة.[4]

> The Prophet said: "Religion is easy and whoever overburdens himself in his religion will not be able to continue in that way. So you should be straight, facilitate understanding and receive the good tidings that you will be rewarded; and gain strength by worshipping in the mornings, the evening and part of the night."

We pray to Allah to keep us on the right path and help us to keep our minds open for ideas that are consistent with new knowledge without contradicting the basic principles of our *dīn*.

Muzammil H. Siddiqi
Chairman, Fiqh Council of North America

INTRODUCTION

The debate about naked-eye moonsighting versus calculation goes back to the first century of the Islamic era. In fact, this debate had taken multiple twists and turns among Jewish scholars centuries before its systematic deliberation by Muslim scholars. Given its long history, many of the arguments on both sides have almost become standardized. The nature of this debate and scholarly reasoning among Muslims are a little different from that of the Jews, for the former have two divinely inspired sources of law that are historically authenticated. The Islamic discourse has predominantly revolved around various Qur'anic verses and the Sunnah. As classical Islamic jurisprudential discourse is heavily dependent upon these two sources, most of the traditional arguments reoccur in the books of *tafsīr*, hadiths, fiqh, and fatwas.

The main arguments against using calculations in determining the beginning or end of Ramadan can be summarized as follows[1]:

1: The Qur'an requires naked-eye sighting of the moon to confirm the beginning and end of Ramadan, Dhu al-Hijjah, and other devotional months. The act of "witnessing the month of Ramadan" stated in the Qur'an means actually sighting the new moon of Ramadan with one's eyes. The Qur'anic term for the crescent moon is *al-hilāl*, which, they argue, literally means "the sighted moon."

2: The prophetic traditions, which reach to the level of infallibility, also require naked-eye sighting in these cases. The prophetic injunctions approve only two definitive methods: naked-eye sighting and completing thirty days.

3: Over the centuries, Islamic scholarship has accepted these two methods as the norm, and thus jurists have rejected all efforts to use mathematically computed astronomical calculations* and opposed those who did so, either in part or in total. Some of these scholars, among them Ibn al-Shikhir, Ibn Surayj, and al-Subki, were otherwise highly

---

★ Throughout this book, I use "calculations" with the understanding that they are "mathematically computed astronomical calculations."

respected and appreciated for their knowledge and piety. Rejecting calculations has been an established norm in all Sunni and Shiʿi jurisprudential schools of thought.

4: The majority of jurists has not advocated for accepting the use of calculations. But a small minority has, based upon some misguided or weak interpretations. Even this minority opinion, however, does not permit bypassing naked-eye sighting altogether, for calculations can be used if the moon is obscured by clouds.

5: Naked-eye sighting has been the universal prophetic practice. The Jews were required to follow this method in determining their lunar months, and history has shown that they adhered to it. Over time, however, they changed this practice in an effort to synchronize their religious festivals and schedules with the solar civil calendars of the secular authorities (e.g., emperors). Devising a calculated calendar for Islamic months, therefore, would be nothing short of following the Jews into misguidance and error.

6: Some contemporary scholars contend that the pre-Islamic Arabs as well as first-generation Muslims were quite capable of employing calculations to determine the lunar months. But the Muslims rejected this method because the Sunnah prohibited it. Astronomy only reached its climax among Muslim scholars, owing to their special interest in monitoring the movements of the sun, moon, planets, and other celestial bodies. Some jurists, among them Shihab al-Din al-Qarrafi and Ibn al-ʿArabi, were also qualified astronomers who could establish precise calendars for lunation, recognize conjunctions, and predict the possible appearance of the new moon. In spite of that, however, they did not use calculations to determine the beginning and end of Ramadan and other Islamic months. As observational astronomy has not developed much over the past centuries, the Fiqh Council's current dependence upon scientifically proven calculations to fix an Islamic calendar is not progress, but sheer backwardness.[2]

I would like to state from the outset that all of these arguments do not stand up to a detailed analysis. The Qur'an never required naked-eye sighting of the moon, but asked Muslims to "witness the month." Scholars agree that the Qur'anic phrase "whoever witnessed the month" means whoever is present in his/her place of residence and comes to learn of

Ramadan's arrival in any way, including (but not confined to) sighting. In reality, the prophetic reports require "certainty" vis-à-vis Ramadan's arrival and end. Physical sighting of the new moon was prescribed as a *means of achieving* that certainty, not the *objective* of fasting. The Qur'anic word *al-ahillah* is the plural of *al-hilāl,* which denotes "the beginning part of something like rain, an announcement, a cry of joy, calling out in a loud voice." *Al-hilāl* has been used both culturally and metaphorically to symbolize the new moon of the first two to seven nights and then the last two nights of the month.

In addition, there are conflicting reports about the Prophet's actual response upon looking at the new moon for the first time. One hadith states that he used to turn his face away from it and seek Allah's protection from its evils. The other popular report indicates that he would recite specific supplications upon sighting it. Al-Bukhari, Muslim, and other hadith authorities never reported any of these supplications; Ibn Dawud stated that both the above-mentioned purported prophetic reports were untrustworthy. The assertion that completing thirty days in the case of obscurities (*ikmāl*) is the only permitted alternative (second to actual sighting) is also debatable.

There is no ijmaᶜ (scholarly consensus) that using calculations to determine Ramadan is prohibited (*ḥarām*), especially if there are obscurities. Jurisprudential literature contains several assertions of such an ijmaᶜ; however, no such thing exists in reality. This debate has been going on since the time of the Successors. The majority of classical jurists opposed the use of calculations for several reasons. There is a long list of possibly valid reasons as to why the classical juristic discourse has been so strongly opposed to this method, especially in matters of faith (*imān*) and worship (*ᶜibādah*). In the past, scholars like Mutarrif ibn ᶜAbd Allah ibn al-Shikhir, Ibn Surayj, Ibn Qutaybah, Taqi al-Din al-Subki, and others actually deployed calculations to confirm or negate Ramadan if there were atmospheric obscurities. Many of their colleagues and later classical jurists rebuked them for doing so, as will be discussed later in this book.

The use of such calculations in religious matters has been dubbed a Jewish innovation, for the fixed Jewish calendar is an intercalated calendar. The rabbis did their best to synchronize their lunar calendar with the solar calendar in order to celebrate religious festivals during specific seasons and to harmonize their holidays with the civil holidays. The pre-Qur'anic Arabs also used to add extra days and an extra thirteenth month in their lunar calendar almost every three years so that hajj would fall during those seasons that were good for travel and business. Thus the sacred period was compromised and hajj was performed at the wrong time. The Prophet insisted upon restoring the calendar and stipulated that each of the new months would start only after the actual sighting of the crescent moon.

Ismail K. Poonawala, author of "Ramadan" in Encarta, explains

> In the pre-Islamic Arabic calendar, the month of Ramadan fell during the heat of summer. The word Ramadan means "scorcher" in Arabic. The early Arabic calendar, like the current Islamic calendar, was lunar. Because a lunar month has only 29 or 30 days, a year of 12 lunar months falls short of the 365 days in a solar calendar. In the pre-Islamic calendar, the lunar months kept their place in the seasons by the insertion of an extra month every two or three years. The Islamic calendar abolished this practice and fixed the Islamic year at 12 months totaling 354 days. As a result Ramadan occurs about 11 days earlier each year, and it rotates through the seasons in a cycle totaling about 33 years.[3]

Therefore, developing and then using precise calculations to determine the new moon's appearance, as well as to confirm or negate the actual new moon of the Islamic lunar months, would not subject the Muslims to imitating the Jews in religious matters. This does not change the sacred period of the established lunar months, but rather helps us determine them with precision and certainty.

Finally, astronomy has made great advances over the centuries. Ancient Babylonian and Greek astronomical practices were based upon an incorrect understanding of the solar system, and Ptolemaic astronomical principles, which had predominated since the second century AC, were also flawed:

> The Ptolemaic theory held that Earth is stationary and at the center of the universe; closest to Earth is the moon, and beyond it, extending outward, are Mercury, Venus, and the sun in a straight line, followed successively by Mars, Jupiter, Saturn, and the so-called fixed stars.[4]

The Muslims were keen observers; a few of them were even proficient astronomers. While they tried to fix many faulty aspects of Ptolemaic astronomy, they did not fully succeed because the Ptolemaic geocentric theory of the universe was based upon an error. According to Fred L. Whipple and Vera C. Rubin:

> After the decline of classical Greek culture Arabian astronomers attempted to perfect the system by adding new epicycles to explain unpredicted variations in the motions and positions of the planets. These efforts failed, however, to resolve the many inconsistencies in the Ptolemaic system, which was finally superseded in the 16th century by the Copernican system.[5]

The systematic explanation that the universe is solocentric, as opposed to geocentric, is credited to Nicolaus Copernicus, who, in 1543, explained the movement of the planets around the sun:

> The Copernican system advanced the theories that the earth and the planets are all revolving in orbits around the sun, and that the earth is spinning on its north–south axis from west to east at the rate of one rotation per day. These two hypotheses superseded the Ptolemaic system, which had been the basis of astronomical theory until that time. Publication of the Copernican system stimulated the study of astronomy and mathematics and laid the basis for the discoveries of the German astronomer Johannes Kepler and the British astronomer Sir Isaac Newton.[6]

Medieval Islamic civilization witnessed a flourishing of scientific development and progress in many fields, including astronomy. But the Muslims' astronomical discoveries were hampered by several innate mistakes in the foundational principles of Ptolemaic theory. Medieval scientists also lacked most of our modern instruments (e.g., high-level telescopes), well-equipped laboratories, NASA's expeditions to the moon, and many other discoveries that have helped astronomy branch out into gamma-ray astronomy (e.g., X-ray, ultraviolet, infrared, radio, and radar astronomy). Countless projects have focused entirely upon the earth– moon relationship and the moon's movement in its orbit around the earth. For instance, since 1957 the IGY (International Geophysical Year) project has used powerful Markowitz cameras, located at over twenty observatories worldwide, to study and then calculate the moon's movements and positions.[7]

CHAPTER 1

THE CULTURAL MILIEU AND ITS *FIQHĪ* RELEVANCE

Human beings are social creatures and thus naturally react to their cultural milieu. Islamic history has witnessed two competing tendencies. I could say that this dispositional variety is part of human nature and is also reflected in the Companions' thinking patterns. Al-Bukhari reports, on the authority of Ibn ʿUmar, that after the battle of Confederates the Prophet commanded some Companions not to pray *ʿaṣr* until they reached the territory of the Banu Qurayzah (a Jewish tribe on the outskirts of Madinah). Some of them, however, prayed it on the way, thinking that the Prophet intended them to reach that spot and did not want them to miss the prayer. The second group followed his command to the letter and prayed only after reaching that spot. The Prophet understood both groups' intention and did not rebuke either of them:

حَدَّثَنا عَبْدُ اللَّهِ بْنُ مُحَمَّدِ بْنِ أَسْمَاءَ قَالَ حَدَّثَنا جُوَيْرِيَةُ عَنْ نَافِعٍ عَنْ ابْنِ عُمَرَ قَالَ قَالَ النَّبِيُّ صَلَّى اللَّهُ عَلَيْهِ وَسَلَّمَ لَنَا لَمَّا رَجَعَ مِنْ الأَحْزَابِ لا يُصَلِّيَنَّ أَحَدٌ الْعَصْرَ إِلا فِي بَنِي قُرَيْظَةَ فَأَدْرَكَ بَعْضَهُمْ الْعَصْرُ فِي الطَّرِيقِ فَقَالَ بَعْضُهُمْ لا نُصَلِّي حَتَّى نَأْتِيَهَا وَقَالَ بَعْضُهُمْ بَلْ نُصَلِّي لَمْ يُرَدْ مِنَّا ذَلِكَ فَذُكِرَ لِلنَّبِيِّ صَلَّى اللَّهُ عَلَيْهِ وَسَلَّمَ فَلَمْ يُعَنِّفْ وَاحِدًا مِنْهُمْ.[1]

Narrated Ibn ʿUmar: On the day of al-Ahzab, (i.e. Clans) the Prophet said, "None of you should affer the *ʿaṣr* prayer but at Banu Qurayzah's place." The *ʿaṣr* prayer became due for some of them on the way. Some of those said, "We will not offer it till we reach Banu Qurayzah" while some others said, "No, we will pray at this spot, for the Prophet did not mean that for us." Later on it was mentioned to the Prophet and he did not berate any of the two groups.

Muslim, on the other hand, reports that the prophetic command specified *ẓuhr:*

و حَدَّثَنِي عَبْدُ اللَّهِ بْنُ مُحَمَّدِ بْنِ أَسْمَاءَ الضُّبَعِيُّ حَدَّثَنا جُوَيْرِيَةُ بْنُ أَسْمَاءَ عَنْ نَافِعٍ عَنْ عَبْدِ اللَّهِ قَالَ نَادَى فِينَا رَسُولُ اللَّهِ صَلَّى اللَّهُ عَلَيْهِ وَسَلَّمَ يَوْمَ انْصَرَفَ عَنْ الأَحْزَابِ أَنْ لا يُصَلِّيَنَّ أَحَدٌ الظُّهْرَ إِلا فِي بَنِي قُرَيْظَةَ فَتَخَوَّفَ نَاسٌ فَوْتَ الْوَقْتِ فَصَلَّوْا دُونَ بَنِي قُرَيْظَةَ وَقَالَ آخَرُونَ لا نُصَلِّي

إلا حَيْثُ أَمَرَنَا رَسُولُ اللَّهِ صَلَّى اللَّهُ عَلَيْهِ وَسَلَّمَ وَإِنْ فَاتَنَا الْوَقْتُ قَالَ فَمَا عَنَّفَ وَاحِدًا مِنْ الْفَرِيقَيْنِ.[2]

ʿAbd Allah reported that the day the Prophet returned from al-Ahzab he announced among us not to offer *ẓuhr* except at Banu Qurayzah's place. Some people prayed before reaching the Banu Qurayzah quarters fearing that the prayer time would be over by that time. The others insisted to pray at Banu Qurayzah, as the Prophet (apparently) commanded, even if the prayer time would have been over. The Prophet did not rebuke any of the two groups.

Muslims have always been concerned that one day the pristine Islamic teachings might be watered down to fit them into the changing realities of time. A group of scholars have been bent upon implementing the Lawgiver's original intention, as they have understood it, in its pristine simplistic formulation. They tend to consider anything outside of that box an innovation designed to compromise Islam's true essence. This tendency, which is not confined to a specific *fiqhī* school, is an ideological movement that sincerely advocates preserving the prophetic Sunnah to the letter. The second tendency struggles to implement the Islamic teachings in their pristine intended forms, though usually not to the letter. They look for some possible reasons behind these prophetic traditions in order to implement them in essence and, whenever possible, to the letter.

The first group insists that the Lawgiver must also detail the specified reasons and the means.

THE DEBATE OVER THE PRAYER TIMINGS AND THE DIRECTION OF THE QIBLAH

The debate over using mathematical calculations to establish the prayer times is a good example of these competing tendencies. The early authorities in most of the known *fiqhī* schools fought against the use of calculations (as contrary the established Sunnah, which required looking at the shadow or the phenomenon of sunrise and sunset for the prayer times). For instance, Ibn Rajab al-Hanbali, rejected the use of calculation:

فتبين أن ديننا لا يحتاج إلى حساب ولا كتاب، كما يفعله أهل الكتاب من ضبط عباداتهم بمسير الشمس وحسباناتها، وأن ديننا في ميقات الصيام معلق بما يرى بالبصر وهو رؤية الهلال، فإن غم أكملنا عدة الشهر و لم نحتج إلى حساب. وإنما علق بالشمس مقدار النهار الذي يجب الصيام فيه، وهو متعلق بأمر مشاهد بالبصر – أيضا –، فأوله طلوع الفجر الثاني، وهو مبدأ ظهور الشمس

على وجه الأرض، وآخره غروب الشمس. كما علق بمسير الشمس أوقات الصلاة، فصلاة الفجر أول وقتها طلوع هذا الفجر، وآخره طلوع الشمس، وأول وقت الظهر زوال الشمس، وآخره مصير ظل كل شيء مثله، وهو أول وقت العصر، وآخره اصفرار الشمس أو غروبها، وهو أول وقت المغرب، وآخره غروب الشفق، وهو أول وقت العشاء، وآخره نصف الليل أو ثلثه، ويمتد وقت أهل الأعذار إلى طلوع الفجر، فهذا كله غير محتاج إلى حساب ولا كتاب.[3]

Therefore, it proves that our religion does not need calculations or writing like the People of the Book who synchronize their acts of worship to the solar movements and calculations. Our religion, in determining the fasting, is dependent upon things seen, such as sighting the crescent moon. We complete thirty days of the month in the case of obscurities and do not need calculations. The obligated (length) day of fasting is also dependent upon the sun. This is also connected with a seen phenomenon. It begins with the dawn… and ends with sunset. Our religion has also connected the prayer timings with the sun. The beginning of *fajr* (morning) is connected with the dawn and its end with the sunrise. The start of *ẓuhr* (noon) prayer is with the meridian and it ends when the shadow equals the actual object… All this does not need either calculation or writing.

Ibn Daqiq al-ʿId, a Shafi`i, also opposed their use:

وَقَالَ ابْنُ دَقِيقِ الْعِيدِ: الْحِسَابُ لا يَجُوزُ الاعْتِمَادُ عَلَيْهِ فِي الصَّلاةِ.[4]

Ibn Daqiq al-ʿId said that calculations cannot be implied in establishing prayer (timings).

Later jurists, such as the Maliki Shihab al-Din al-Qarrafi, strongly opposed this tendency (of refuting calculation in fixing prayer times). In fact, al-Qarrafi's belief that calculations could be used to determine the prayer times was a big leap forward. His allowance of this practice should be understood in the context of his heated dialogue with his opponents.

How many scholars presently maintain that using calculations to determine the prayer times is unnecessary and forbidden? Muhammad Rashid Rida observes:

إِنَّ إِثْبَاتَ أَوَّلِ شَهْرِ رَمَضَانَ وَأَوَّلِ شَهْرِ شَوَّالٍ هُوَ كَإِثْبَاتِ أَوْقَاتِ الصَّلَوَاتِ الْخَمْسِ قَدْ نَاطَهَا الشَّارِعُ كُلَّهَا بِمَا يَسْهُلُ الْعِلْمُ بِهِ عَلَى الْبَدْوِ وَالْحَضَرِ... وَغَرَضُ الشَّارِعِ مِنْ ذَلِكَ الْعِلْمُ بِهَذِهِ الأَوْقَاتِ لا التَّعَبُّدُ بِرُؤْيَةِ الْهِلالِ وَلا بِتَبَيُّنِ الْخَيْطِ الأَبْيَضِ مِنَ الْخَيْطِ الأَسْوَدِ مِنَ الْفَجْرِ... وَلا التَّعَبُّدُ بِرُؤْيَةِ ظِلِّ الزَّوَالِ وَقْتَ الظُّهْرِ ، وَصَيْرُورَةِ ظِلِّ الشَّيْءِ مِثْلَهُ وَقْتَ الْعَصْرِ ، وَلا بِرُؤْيَةِ غُرُوبِ

الشَّمْسِ وَغِيبَةِ الشَّفَقِ لِوَقْتَي الْعِشَاءَيْنِ ، فَغَرَضُ الشَّارِعِ مِنْ مَوَاقِيتِ الْعِبَادَةِ مَعْرِفَتُهَا وَمَا ذَكَرَهُ - صَلَّى اللَّهُ عَلَيْهِ وَسَلَّمَ - مِنْ نَوْطِ إِثْبَاتِ الشَّهْرِ بِرُؤْيَةِ الْهِلالِ أَوْ إِكْمَالِ الْعِدَّةِ بِشَرْطِهِ قَدْ عَلَّلَهُ بِكَوْنِ الأُمَّةِ فِي عَهْدِهِ كَانَتْ أُمِّيَّةً أَوْ مِنْ مَقَاصِدِ بِعْثَتِهِ إِخْرَاجُهَا مِنَ الأُمِّيَّةِ لا إِبْقَاؤُهَا.[5]

Determining start of the months of Ramadan and Shawwal is just like determining timings of the five times daily prayers. The Lawgiver has allowed that determination to take place with whatever source of knowledge is easily available to the villagers as well as town people... The Lawgiver's intent is the knowledge of these timings and not worship through moon sighting or differentiation of white thread from the black thread at dawn. Likewise there is no worship involved in seeing the shadow at the noon time for *ẓuhr* or shadow becoming equal to its object for *ʿaṣr* or seeing the sunset or twilight for *maghrib* and *ʿishāʾ*. The intent of the Lawgiver is just identification of these exact timings [and not worship of these timings or the means or process involved]. The Prophet had connected confirmation of the month by actual moon sighting or completion with the condition that the Ummah was unlettered during his times. The goal of his prophethood has been to bring the Ummah out of its unlettered status and not to continue sustaining it.

He continues:

وَجُمْلَةُ الْقَوْلِ أَنَّنَا بَيْنَ أَمْرَيْنِ : إِمَّا أَنْ نَعْمَلَ بِالرُّؤْيَةِ فِي جَمِيعِ مَوَاقِيتِ الْعِبَادَاتِ أَخْذًا بِظَوَاهِرِ النُّصُوصِ وَحُسْبَانِهَا تَعَبُّدِيَّةً ، وَحِينَئِذٍ يَجِبُ عَلَى كُلِّ مُؤَذِّنٍ ألا يُؤَذِّنَ حَتَّى يَرَى نُورَ الْفَجْرِ الصَّادِقِ مُسْتَطِيرًا مُنْتَشِرًا فِي الأُفُقِ ، وَحَتَّى يَرَى الزَّوَالَ وَالْغُرُوبَ إِلَخْ ، وَإِمَّا أَنْ نَعْمَلَ بِالْحِسَابِ الْمَقْطُوعِ بِهِ لأَنَّهُ أَقْرَبُ إِلَى مَقْصِدِ الشَّارِعِ ، وَهُوَ الْعِلْمُ الْقَطْعِيُّ بِالْمَوَاقِيتِ وَعَدَمِ الاخْتِلافِ فِيهَا... وَأَمَّا هَذَا الاخْتِلافُ وَتَرْكُ النُّصُوصِ فِي جَمِيعِ الْمَوَاقِيتِ - عَمَلا بِالْحِسَابِ مَا عَدَا مَسْأَلَةَ الْهِلالِ - فَلا وَجْهَ وَلا دَلِيلَ عَلَيْهِ ، وَلَمْ يَقُلْ بِهِ إِمَامٌ مُجْتَهِدٌ بَلْ هُوَ مِنْ قَبِيلِ (أَفَتُؤْمِنُونَ بِبَعْضِ الْكِتَابِ وَتَكْفُرُونَ بِبَعْضٍ) (البقرة: 85).[6]

In short we are left with only two options. Either we take all these texts literally and go by actual sighting in determining all times connected with acts of worship. That will be required if we take this sighting as an act of worship in itself. This will mandate all callers of *adhān* not to call *adhān* until they actually sight the early morning light spreading over the horizon or sight the noon or sunset [by their own eyes] etc. Or we can adopt the precisely categorical calculations as they are closer to the objectives of the Lawgiver. The Lawgiver's true intent [through these texts] is to identify the precise knowledge of these timings (of acts of worship) so that there is a unity [and not disunity among believers] in this regard... There are no legal basis nor any solid proof for this disparity of adopting the calculations to determine all sacred timings and discard them only in confirming the new moon. No religious authority has maintained that. In reality it is a reflection of what the Qur'an states in Surah 2:85, "Do you believe in some portions of the Book and deny the others?"

The early jurists also argued that the qiblah lay between the East and the West. Initially, when some Muslim scholars endeavored to specify its exact direction with the help of mathematically computed astronomical calculations, they were scolded by the leading authorities in hadith, fiqh, and other Islamic fields. Space does not permit a detailed account of the evolution that has occurred in many *fiqhī* areas over the centuries, so I will confine myself to only a few examples.

For instance, Ibn Rajab al-Hanbali opposed the use of calculations to determine the qiblah. Following the lead of Ahmad ibn Hanbal, he prohibited this practice:

وكذلك القبلة، لا تحتاج إلى حساب ولا كتاب، وإنما تعرف في المدينة وما سامتها من الشام والعراق وخراسان بما بين المشرق والمغرب.[7]

The [direction of the] qiblah does not need either writing or calculation. The qiblah is known by the city [location] and its comparison with Syria, Iraq, and Khurasan, given that it is between the East and the West.

ونقل الأثرم، عن أحمد، أنه قيل له: قبلة أهل بغداد على الجدي؟ فجعل ينكر أمر الجدي، فقال: أيش الجدي؟ ولكن على حديث عمر: (ما بين المشرق والمغرب قبلة).[8]

Reported al-Athram that someone said to Imam Ahmad that the [direction of the] qiblah from Baghdad was toward the North Star. Ahmad, refuting this North Star basis, said, "What is [this] North Star? [The qiblah of Baghdad is not based upon the North Star] but upon the hadith of ʿUmar, 'The qiblah is between the East and the West.'"

Like countless other jurists, Ibn Rajab thought that demanding precision in deciding the qiblah and fixing the prayer times with the help of calculations was tantamount to degrading the Companions' prayers, for they neither determined the precise qiblah directions from the distant conquered cities nor used calculations to determine the prayer times.

وهذا يفضي إلى تضليل سلف الأمة، والطعن في صلاتهم.[9]

This leads to [saying] that the worthy ancestors of the Ummah were misguided and also to slandering their prayers.

Currently, how many scholars would argue that employing astronomical calculations in determining the qiblah is tantamount to slandering the prayers of our noble ancestors?

THE DEBATE ABOUT THE UNITY OR VARIETY OF HORIZONS

Early Hanafi, Maliki, Hanbali, and some Shafiʿi jurists, strong believers in the unity of horizons, imposed fasting on the entire Ummah when the new moon was sighted in any Muslim city or country. The earlier jurists even asserted that this fasting was required by the prophetic command itself.

ذَهَبَ الْحَنَفِيَّةُ وَالْمَالِكِيَّةُ وَالْحَنَابِلَةُ وَهُوَ قَوْلٌ عِنْدَ الشَّافِعِيَّةِ: إِلَى عَدَمِ اعْتِبَارِ اخْتِلافِ الْمَطَالِعِ فِي إِثْبَاتِ شَهْرِ رَمَضَانَ ، فَإِذَا ثَبَتَ رُؤْيَةُ هِلالِ رَمَضَانَ فِي بَلَدٍ لَزِمَ الصَّوْمُ جَمِيعَ الْمُسْلِمِينَ فِي جَمِيعِ الْبِلادِ ، وَذَلِكَ لِقَوْلِهِ صلى الله عليه وسلم: (صُومُوا لِرُؤْيَتِهِ) وَهُوَ خِطَابٌ لِلأُمَّةِ كَافَّةً.[10]

According to Hanafi, Maliki, Hanbali (and according to one report from the Shafiʿi school of thought), no consideration is given to the diversity of horizons in regard to confirming the month of Ramadan. The entire Muslim world is required to begin fasting if the new moon is sighted anywhere in the world. This is in line with the prophetic tradition "start fasting by seeing the new moon." The hadith is addressed to the entire Muslim nation.

Ibn Qudamah reports the Hanbali opinion of the unity of horizons:

وأجمع المسلمون على وجوب صوم شهر رمضان وقد ثبت أن هذا اليوم من شهر رمضان بشهادة الثقات فوجب صومه على جميع المسلمين ولأن شهر رمضان ما بين الهلالين ... فيجب صيامه بالنص والإجماع ولأن البينة العادلة شهدت برؤية الهلال فيجب الصوم.[11]

Ramadan is an agreed upon obligation. It was proven by the trustworthy witnesses that the day was the month of Ramadan. Therefore, it became incumbent upon all the Muslims to fast. The month of Ramadan is between two crescent moons. It must be fasted as is proven by the text and by the consensus because the trustworthy testifier testified the sighting of the new moon. Therefore fasting became obligatory.

Al-Qarrafi explains the Maliki and Hanbali positions:

بِأَنَّ رُؤْيَةَ الْهِلالِ بِمَكَانٍ قَرِيبًا كَانَ أَوْ بَعِيدًا إِذَا ثَبَتَتْ لَزِمَ النَّاسَ كُلَّهُمْ الصَّوْمُ ، وَأَنَّ حُكْمَ مَنْ لَمْ يَرَهُ حُكْمُ مَنْ رَآهُ ، وَلَوْ اخْتَلَفَتْ الْمَطَالِعُ نَصًّا قَالَ أَحْمَدُ الزَّوَالُ فِي الدُّنْيَا وَاحِدٌ بِقَوْلِهِ صلى الله تعالى عليه وسلم (صُومُوا لِرُؤْيَتِهِ) وَهُوَ خِطَابٌ لِلأُمَّةِ كَافَّةً.[12]

The Muslim world in its entirety is required to commence fasting by the report of sighting anywhere in the world, whether the place of sighting is close to or far from them. Those who have not seen the new moon come under the ruling of those who have seen it, even if the horizons are different. Ahmad said that the noon (*zawāl*) all over the world is same as the prophetic commandment states: "start fasting by seeing it." It is directed to the entire Muslim community.

The Hanafi ʿUthman ibn ʿAli al-Zayla'i states:

وَإِذَا ثَبتَ فِي مِصْرَ لَزِمَ سَائِرَ النَّاسِ فَيَلْزَمُ أَهْلَ الْمَشْرِقِ بِرُؤْيَةِ أَهْلِ الْمَغْرِبِ فِي ظَاهِرِ الْمَذْهَبِ.[13]

The authorized position of the school is that the sighting in one city is the sighting for all. It will become incumbent upon the inhabitants of the Eastern hemisphere to confirm the month by a sighting in the West.

وَأَكْثَرُ الْمَشَايِخِ عَلَى أَنَّهُ لا يُعْتَبَرُ حَتَّى إذَا صَامَ أَهْلُ بَلْدَةٍ ثَلاثِينَ يَوْمًا وَأَهْلُ بَلْدَةٍ أُخْرَى تِسْعَةً
وَعِشْرِينَ يَوْمًا يَجِبُ عَلَيْهِمْ قَضَاءُ يَوْمٍ.[14]

The majority of [Hanafi] elders give no consideration [to the diversity of horizons], so much so that they require making up for a day of fasting if people of one locality fasted for thirty days and the other locality for twenty-nine days.

The Maliki Ibn Abi Zayd al-Qayrawani connects fasting with sighting the moon for all the Muslims.

وَصَوْمُ شَهْرِ رَمَضَانَ فَرِيضَةٌ يُصَامُ لِرُؤْيَةِ الْهِلَالِ وَيُفْطَرُ لِرُؤْيَتِهِ.[15]

Fasting the month of Ramadan is obligatory. Fasting begins (for all) with moon sighting and ends with moon sighting.

Imam al-Shawkani reports that the Zaydi school of thought also maintained global sighting.

وَالَّذِي يَنْبَغِي اعْتِمَادُهُ هُوَ مَا ذَهَبَ إِلَيْهِ الْمَالِكِيَّةُ وَجَمَاعَةٌ مِنْ الزَّيْدِيَّةِ. وَاخْتَارَهُ الْمَهْدِيُّ مِنْهُمْ
وَحَكَاهُ الْقُرْطُبِيُّ عَنْ شُيُوخِهِ أَنَّهُ إذَا رَآهُ أَهْلُ بَلَدٍ لَزِمَ أَهْلَ الْبِلادِ كُلِّهَا.[16]

The Maliki and Zaydi position should be accepted in this regard. Al-Mahdi and al-Qurtabi have also reported on the authority of their elders that the sighting in one locality is the sighting of all the localities.

These early jurists interpreted the hadith "start fasting by sighting it and stop fasting by sighting it" as a universal command for all Muslims everywhere to fast during Ramadan from a trustworthy sighting in any locality. Interestingly, they emphasized the point that local sighting was a legal reason (*sabab sharʿī*) for all Muslims to begin fasting. For instance, al-Qarrafi states the Maliki and Hanbali position:

أَنَّ الْمَالِكِيَّةَ جَعَلُوا رُؤْيَةَ الْهِلالِ فِي بَلَدٍ مِنْ الْبِلادِ سَبَبًا لِوُجُوبِ الصَّوْمِ عَلَى جَمِيعِ أَقْطَارِ الأَرْضِ وَوَافَقَتْهُمْ الْحَنَابِلَةُ رَحِمَهُمُ اللَّهُ عَلَى ذَلِكَ. وَقَالَتْ الشَّافِعِيَّةُ رَحِمَهُمُ اللَّهُ لِكُلِّ قَوْمٍ رُؤْيَتُهُمْ وَاتَّفَقَ الْجَمِيعُ عَلَى أَنَّ لِكُلِّ قَوْمٍ فَجْرَهُمْ وَزَوَالَهُمْ وَعَصْرَهُمْ وَمَغْرِبَهُمْ وَعِشَاءَهُمْ.[17]

The Malikis made the local sighting of the moon sighting a legal cause for the entire world to fast and the Hanbalis agreed to that. The Shafiʿis stated that each locality has its own local sighting. All of these scholars agreed that each locality has its own *fajr*, *ẓuhr*, *ʿaṣr*, *maghrib*, and *ʿishāʾ* timings.

In spite of these strong early positions, however, later jurists rejected the unity of horizon positions officially adopted by their schools.

Al-Qarrafi, another Maliki authority, proved with the help of astronomy that the horizons are different. He argued that people in the Western Hemisphere have more chances of sighting the new moon even if the people in the East cannot see it. The newly born moon increases in age and volume as the sun moves in a westerly direction. The deflection of light through the moon increases with the moon's gradual movement from its conjunction point and with its age after sunset.

وَبَيَّنَ الْقَرَافِيُّ اخْتِلافَ مَطَالِعِ الْهِلالِ عِلْمِيًّا ، وَذَكَرَ سَبَبًا مِنْ أَسْبَابِهِ مُكْتَفِيًا بِهِ عَنْ الْبَقِيَّةِ الْمَذْكُورَةِ فِي عِلْمِ الْهَيْئَةِ: وَهُوَ أَنَّ الْبِلادَ الْمَشْرِقِيَّةَ إِذَا كَانَ الْهِلالُ فِيهَا فِي الشُّعَاعِ وَبَقِيَتْ الشَّمْسُ تَتَحَرَّكُ مَعَ الْقَمَرِ إِلَى الْجِهَةِ الْغَرْبِيَّةِ فَمَا تَصِلُ الشَّمْسُ إِلَى أُفُقِ الْمَغْرِبِ إلا وَقَدْ خَرَجَ الْهِلالُ عَنْ الشُّعَاعِ فَيَرَاهُ أَهْلُ الْمَغْرِبِ وَلا يَرَاهُ أَهْلُ الْمَشْرِقِ.[18]

Al-Qarrafi also established that the prayer times are different in different localities. It is agreed that if there were two brothers, one living in the East and other living in the West, and that both of them died at noon, the latter would inherit from the former because the one in the East had died earlier, since noon in the East arrives earlier than in the West.

إِذَا مَاتَ أَخَوَانِ عِنْدَ الزَّوَالِ أَحَدُهُمَا بِالْمَشْرِقِ وَالآخَرُ بِالْمَغْرِبِ حُكِمَ بِأَسْبَقِيَّةِ مَوْتِ الْمَشْرِقِيِّ؛ لِأَنَّ زَوَالَ الْمَشْرِقِ مُتَقَدِّمٌ عَلَى زَوَالِ الْمَغْرِبِ فَيَرِثُ الْمَغْرِبِيُّ الْمَشْرِقِيَّ.[19]

He concludes:

وَإِذَا كَانَ الْهِلَالُ يَخْتَلِفُ بِاخْتِلَافِ الآفَاقِ وَجَبَ أَنْ يَكُونَ لِكُلِّ قَوْمٍ رُؤْيَتُهُمْ فِي الاهِلَّةِ كَمَا أَنَّ لِكُلِّ قَوْمٍ فَجْرَهُمْ وَغَيْرَ ذَلِكَ مِنْ أَوْقَاتِ الصَّلَوَاتِ، وَهَذَا حَقٌّ ظَاهِرٌ وَصَوَابٌ مُتَعَيِّنٌ أَمَّا وُجُوبُ الصَّوْمِ عَلَى جَمِيعِ الأَقَالِيمِ بِرُؤْيَةِ الْهِلَالِ بِقُطْرٍ مِنْهَا فَبَعِيدٌ عَنِ الْقَوَاعِدِ، وَالأَدِلَّةُ لَمْ تَقْتَضِ ذَلِكَ فَاعْلَمْهُ.[20]

If the crescent moon differs according to the difference in horizons, then it becomes incumbent that each locality must follow its own [sighting of the] moon. So, each locality has its own prayer timings such as *fajr* [prayer]. This is a clear-cut truth and an exactly righteous position. On the other hand, imposing global fasting with a local sighting of the moon is far from the [established Islamic] rules. One needs to know that the [Islamic] texts did not require such an obligation.

Al-Nawawi, a Shafiʿi authority, observes:

إِذَا رَأَوُا الْهِلَالَ فِي رَمَضَانَ فِي بَلَدٍ وَلَمْ يَرَوْهُ فِي غَيْرِهِ، فَإِنْ تَقَارَبَ الْبَلَدَانِ فَحُكْمُهُمَا بَلَدٌ وَاحِدٌ وَيَلْزَمُ أَهْلَ الْبَلَدِ الاخَرِ الصَّوْمُ بِلَا خِلَافٍ وَإِنْ تَبَاعَدَا فَوَجْهَانِ مَشْهُورَانِ فِي الطَّرِيقَتَيْنِ (أَصَحُّهُمَا) لَا يَجِبُ الصَّوْمُ عَلَى أَهْلِ الْبَلَدِ الآخَرِ، وَبِهَذَا قَطَعَ الْمُصَنِّفُ وَالشَّيْخُ أَبُو حَامِدٍ وَالْبُنْدَنِيجِيّ وَآخَرُونَ، وَصَحَّحَهُ الْعَبْدَرِيُّ وَالرَّافِعِيُّ وَالأَكْثَرُونَ. (وَالثَّانِي) يَجِبُ وَبِهِ قَالَ الصَّيْمَرِيُّ وَصَحَّحَهُ الْقَاضِي أَبُو الطَّيِّبِ وَالدَّارِمِيُّ وَأَبُو عَلِيٍّ السِّنْجِيُّ وَغَيْرُهُمْ.[21]

If the moon is sighted in one locality and not seen in the others, then the sighting will be applicable to the nearby areas only. There is no difference of opinion regarding this issue. The difference is only regarding the distant localities. There are two known opinions about this matter. The correct one of the two opinions is that people are not required to follow the sighting of another locality. This is what al-Musannif, Shaykh Abu Hamid, al-Bandaniji, and others have categorically established. Many others like al-Abdari and al-Rafaʿ have verified the same. On the other hand, others such as al-Sumairi, al-Qadi Abu al-Tayyib, al-Darimi, and Abu ʿAli al-Sanji have ruled that the sighting of one locality applies to all.

In addition, he reports that:

وَنَقَلَ ابْنُ الْمُنْذِرِ عَنْ عِكْرِمَةَ وَالْقَاسِمِ وَسَالِمٍ وَإِسْحَاقَ بْنِ رَاهَوَيْهِ أَنَّهُ لَا يَلْزَمُ غَيْرَ أَهْلِ بَلَدِ الرُّؤْيَةِ.[22]

Ibn al-Munzir has narrated that ʿIkrimah, al-Qasim, Salim, and Ishaq ibn Rahawayh all agreed that a sighting in one locality does not apply to the other localities.

Among the Hanafi scholars, authorities such as al-Zayla'i have argued for local sighting.

وَقِيلَ يَخْتَلِفُ بِاخْتِلافِ الْمَطَالِعِ لأَنَّ السَّبَبَ الشَّهْرُ وَانْعِقَادُهُ فِي حَقِّ قَوْمٍ لِرُؤْيَةٍ لا يَسْتَلْزِمُ انْعِقَادَهُ فِي حَقِّ آخَرِينَ مَعَ اخْتِلافِ الْمَطَالِعِ وَصَارَ كَمَا لَوْ زَالَتْ أَوْ غَرَبَتْ الشَّمْسُ عَلَى قَوْمٍ دُونَ آخَرِينَ وَجَبَ عَلَى الأَوَّلِينَ الظُّهْرُ وَالْمَغْرِبُ دُونَ أُولَئِكَ.[23]

Unlike the official position, some of the jurists have adopted the local sighting [instead of universal sighting] because the month [of Ramadan] is confirmed by sighting. The sighting in one locality cannot be applied to the other areas with different horizons. It is just like the noon and sunset timings. The sunset timings in one locality would not require the *maghrib* prayer at the other locality.

Ibn ʿAbidin argues that the fact that horizons are different cannot be debated. He also talks about the difference in the noon and sunset timings at different localities. Every second, given the sun's westward movement, it is morning for a new locality and evening for another.[24]

Al-Zayla'i, who prefers local sighting for the same reasons,[25] quotes the following incident to validate his view:

وَرُوِيَ أَنَّ أَبَا مُوسَى الضَّرِيرَ الْفَقِيهَ صَاحِبَ الْمُخْتَصَرِ قَدِمَ الاسْكَنْدَرِيَّة فَسُئِلَ عَمَّنْ صَعِدَ عَلَى مَنَارَةِ الاسْكَنْدَرِيَّةِ فَيَرَى الشَّمْسَ بِزَمَانٍ طَوِيلٍ بَعْدَمَا غَرَبَتْ عِنْدَهُمْ فِي الْبَلَدِ أَيَحِلُّ لَهُ أَنْ يُفْطِرَ فَقَالَ لا وَيَحِلُّ لأَهْلِ الْبَلَدِ لأَنَّ كُلا مُخَاطَبٌ بِمَا عِنْدَهُ وَالدَّلِيلُ عَلَى اعْتِبَارِ الْمَطَالِعِ مَا رُوِيَ عَنْ كُرَيْبٌ.[26]

It is narrated that the jurist Abu Musa al-Darir, the author of *Al-Mukhtaṣar*, came to Alexandria. He was asked about a person who ascended the minaret and continued looking at the sun long after the sun had set for the people on the ground. Was he allowed to break the fast? He answered, "No," whereas the people on the ground could break the fast, because everyone is required to follow his own situation. The report of Kurayb is the fundamental evidence of the local sighting.

Note that both groups cite Kurayb's hadith to draw fundamentally opposite conclusions. This hadith is reported by Imam Muslim:

وَفِي الصَّحِيحِ عَنْ كُرَيْبٍ ، (أَنَّ أُمَّ الْفَضْلِ بَعَثَتْهُ إِلَى مُعَاوِيَةَ بْنِ أَبِي سُفْيَانَ بِالشَّامِ قَالَ: فَقَدِمْتُ الشَّامَ فَقَضَيْتُ حَاجَتَهَا ، وَاسْتَهَلَّ عَلَيَّ هِلالُ رَمَضَانَ وَأَنَا بِالشَّامِ ، فَرَأَيْتُ الْهِلالَ لَيْلَةَ الْجُمُعَةِ، ثُمَّ قَدِمْتُ

الْمَدِينَةَ فِي آخِرِ الشَّهْرِ ، فَسَأَلَنِي ابْنُ عَبَّاسٍ ، ثُمَّ ذَكَرَ الْهِلالَ فَقَالَ: مَتَى رَأَيْتُمُوهُ؟ فَقُلْتُ: لَيْلَةَ الْجُمُعَةِ، فَقَالَ: أَنْتَ رَأَيْتَهُ؟ قُلْتُ: نَعَمْ، وَرَآهُ النَّاسُ وَصَامُوا وَصَامَ مُعَاوِيَةُ قَالَ: لَكِنَّا رَأَيْنَاهُ لَيْلَةَ السَّبْتِ، فَقُلْتُ لَهُ: أَوَلا تَكْتَفِي بِرُؤْيَةِ مُعَاوِيَةَ؟ قَالَ: لا؛ هَكَذَا أَمَرَنَا رَسُولُ اللَّهِ صلى الله عليه وسلم).[27]

Imam Muslim reports that Umm al-Fadl sent Kurayb to Syria to Muʿawiyah to deal with a particular matter. He [Kurayb] said, "I was in Syria when the month of Ramadan was confirmed. I saw the new moon on Friday evening. I returned to Madinah by the end of the month. Discussing the new moon, Ibn ʿAbbas asked me: 'When did you see the new moon?' I replied, 'On Friday evening.' He asked me whether I had seen the moon by myself and I answered in the positive. I also told him that Muʿawiyah along with a multitude of people saw it and confirmed the month accordingly. Ibn ʿAbbas observed that they had actually seen it on Saturday evening. I inquired whether Muʿawiyah's sighting was not adequate [for Ibn ʿAbbas and the people of Madinah]. Ibn ʿAbbas replied in the negative and said, 'This is what the Messenger of Allah had commanded us.'"

The basic argument here is that Ibn ʿAbbas does not accept the sighting in Damascus as applicable to the people in Madinah, as al-Azimabadi summarizes:

ووجه الاحتجاج به أن ابن عباس لم يعمل برؤية أهل الشام وقال في آخر الحديث هكذا أمرنا، فدل ذلك على أنه قد حفظ من رسول الله صلى الله عليه وسلم أنه لا يلزم أهل بلد العمل برؤية أهل بلد آخر.[28]

The argument is based upon the fact that Ibn ʿAbbas did not act upon the report of the sighting from Syria. Ibn ʿAbbas in conclusion said that this was what the Prophet had commanded us. This proves that the Prophet did not require Muslims to follow the report of a sighting in other localities and that was what Ibn ʿAbbas had narrated from the Prophet.

Ibn al-ʿArabi argues that this report validates local sighting and that the horizons are different. Thus, sighting in one area cannot serve as sighting for everyone, especially for those living in distant lands.[29] Al-Bayhaqi, on the other hand, disagrees and states that this report does not vindicate the position that horizons are multiple; rather, it substantiates that the horizons are united.

يَحْتَمِلُ أَنْ يَكُونَ ابْنُ عَبَّاسٍ إِنَّمَا قَالَ ذَلِكَ لانْفِرَادِ كُرَيْبٌ بِهَذَا الْخَبَرِ، وَجَعَلَ طَرِيقَهُ طَرِيقَ الشَّهَادَاتِ، فَلَمْ يَقْبَلْ فِيهِ قَوْلَ الْوَاحِدِ، وَيَحْتَمِلُ أَنْ يَكُونَ قَوْلُهُ: هَكَذَا أَمَرَنَا رَسُولُ اللَّهِ صلى الله

عليه وسلم اعْتِبَارًا بِقَوْلِهِ عليه السلام: (فَإِنْ غُمَّ عَلَيْكُمْ فَأَكْمَلُوا الْعِدَّةَ)، وَيَكُونَ ذَلِكَ قَوْلَهُ، لا فَتْوَى مِنْ جِهَتِهِ، أَخْذًا بِهَذَا الْخَبَرِ.[30]

It is possible that Ibn ʿAbbas rejected that because Kurayb was a single person who witnessed the sighting. Ibn ʿAbbas applied the rule of witnesses, according to which the witness of a single person is not accepted. It is also possible that by his statement "this is what the Messenger of Allah commanded us," Ibn ʿAbbas meant what the Prophet had said elsewhere: "Complete the term [of thirty days] if it is cloudy." Then it would reflect the statement of the Prophet, not the verdict of Ibn ʿAbbas himself.

Ibn Qudamah maintains that this report was rejected because the witness of one person is not enough to confirm the arrival of Ramadan.

فأما حديث كريب فإنما دل على أنهم لا يفطرون بقول كريب وحده.[31]

Al-Shawkani says the same:

وَاعْلَمْ أَنَّ الْحُجَّةَ إِنَّمَا هِيَ فِي الْمَرْفُوعِ مِنْ رِوَايَةِ ابْنِ عَبَّاسٍ لا فِي اجْتِهَادِهِ الَّذِي فَهِمَ عَنْهُ النَّاسُ وَالْمُشَارُ إِلَيْهِ بِقَوْلِهِ: "هَكَذَا أَمَرَنَا رَسُولُ اللَّهِ صلى الله عليه وسلم" هُوَ قَوْلُهُ: فَلا نَزَالُ نَصُومُ حَتَّى نُكْمِلَ ثَلاثِينَ، وَالامْرُ الْكَائِنُ مِنْ رَسُولِ اللَّهِ صلى الله عليه وسلم هُوَ مَا أَخْرَجَهُ الشَّيْخَانِ وَغَيْرُهُمَا بِلَفْظِ: (لا تَصُومُوا حَتَّى تَرَوْا الْهِلالَ، وَلا تُفْطِرُوا حَتَّى تَرَوْهُ فَإِنْ غُمَّ عَلَيْكُمْ فَأَكْمِلُوا الْعِدَّةَ ثَلاثِينَ) وَهَذَا لا يَخْتَصُّ بِأَهْلِ نَاحِيَةٍ عَلَى جِهَةِ الانْفِرَادِ بَلْ هُوَ خِطَابٌ لِكُلِّ مَنْ يَصْلُحُ لَهُ مِنْ الْمُسْلِمِينَ فَالاسْتِدْلالُ بِهِ عَلَى لُزُومِ رُؤْيَةِ أَهْلِ بَلَدٍ لِغَيْرِهِمْ مِنْ أَهْلِ الْبِلادِ أَظْهَرُ مِنْ الاسْتِدْلالِ بِهِ عَلَى عَدَمِ اللُّزُومِ لأَنَّهُ إِذَا رَآهُ أَهْلُ بَلَدٍ فَقَدْ رَآهُ الْمُسْلِمُونَ فَيَلْزَمُ غَيْرَهُمْ مَا لَزِمَهُمْ.[32]

It should be known that the evidence in this report depends upon the fact that Ibn ʿAbbas attributes the judgment to the Prophet and not to himself, by saying "this is what the Prophet has commanded us." The evidence does not lie in Ibn ʿAbbas' ijtihad, as some people have mistakenly understood it, but in his reference to the other prophetic report that says, "We will continue fasting until we complete thirty days." This prophetic tradition is reported by al-Bukhari and Muslim in the following words. "Do not start fasting until you see the new moon and do not stop fasting until you see the new moon. Complete counting thirty days if it is cloudy." This commandment is not confined to one locality but is a generic address to all the eligible Muslims. There is stronger evidence in the report of Kurayb that the people should be required to observe fasting with the report of sighting from anywhere rather than arguing that it substantiates the local sighting. This is because the sighting by one locality is actually the sighting by all the Muslims and the same rule will apply to all of them.

Ibn Qudamah also argues that it substantiates universal sighting.

أن الحديث دل على صحة الوجه الآخر.[33]

The hadith proves authenticity of other opinion (global moon sighting).

Now, what is the truth of this matter? Both learned groups of jurists are asserting varying and conflicting calims based upon the same prophetic hadiths. The Prophet's words and commands cannot mislead jurists in the same school of thought, for they usually follow the same jurisprudential methodology of derivation. This fact can be reconciled only if we accept that each group of scholars is trying to understand and implement the prophetic commands to the best of its ability and in accordance with its cultural milieu. The legal reason for fasting (global sighting), which most of the righteous ancestors (*Salaf*) accepted, was not accepted as the reason by the later jurists of the same schools.

The use of calculations to determine the qiblah, condemned as illegitimate by many leading *fiqhī* authorities, was accepted as very legitimate by those who came later. Likewise, the use of mathematical calculations, dubbed as slandering the Companions' prayers by many early jurists, was welcomed by later jurists. It seems that early jurists interpreted these prophetic hadiths in accordance with their circumstances and with the scientific knowledge available to them, and that later jurists did the same according to the geographical and scientific information available to them.

Given this reality, we can say that these prophetic hadiths are unequivocal in their historical authenticity but not in their interpretations. The Prophet used prophetic terms that allowed a variety of interpretations for people of different times, localities, and ages. This is the miracle of the Islamic texts, namely, that they are suitable for all times and all possible circumstances. Any selective interpretation and insistence upon a narrow implication would limit their infinite wisdom.

AN ANALYSIS OF THE SELECTIVE INSISTENCE AND THE ISSUE OF *TA'WĪL*

I am not accusing early jurists of doing injustice to the prophetic hadiths by fixing their narrow interpretations. Their selective insistence was based upon a variety of external historical reasons, for they were faced with a number of challenges. One of the leading challenges for the first and second generation Islamic scholars was to encounter the extreme literalist and narrow approaches of the Kharijites and Hashwites on the one hand and the metaphorical and allegorical approaches (*ta'wīl*) of the Jahimites and Muᶜtazilites on the other. They also had to face the harsh realities of the Muᶜtazilites' speculative theological interpretations, which rendered many of the Islamic texts almost meaningless. The above-mentioned scholars, among them the four founders of the Sunni legal schools of thought, defended the spirit of the Qur'an and Sunnah to its letter in an effort to stop the possible evils that might arise from allegorical interpretations of the Islamic texts, which had already played havoc with Jewish and Christian texts. The original Biblical text was subjected to extreme violence by classical Jewish and Christian scholarship to such a degree that it lost its meanings and integrity due to the allegorization of its text.

ALLEGORICAL INTERPRETATIONS AND CHRISTIANITY

For instance, the early Church Fathers were quite aware of the incongruity and strangeness inherent in the Biblical texts. "For as someone," observes the early Church Father Origen of Alexandria,

> points out to us the stories of Lot's daughters and their apparently unlawful intercourse with their father, or of Abraham's two wives, or of two sisters who married Jacob, or the two maidservants who increased the number of his sons, what else can we answer than that these are certain mysteries and types of spiritual matters, but that we do not know of what sort they are?[34]

Atheist Greek scholars like Celsus and Porphyry pointed out such immoralities prevalent in the Old Testament[35] as well as quoted several such passages to argue about the Old Testament's human origins. The Church Fathers, on the other hand, could not declare the Old Testament to be man-made and unauthoritative because they believed, as Origen observed, that "the sacred Scriptures were not composed by any human words but were written by the inspiration of the Holy Spirit and were also delivered and entrusted to us by the will of God the Father through His Only Begotten Son Jesus Christ."[36] So it was the normative scripture that, as they viewed it, Jesus followed and urged others to regard as the key to understanding his person. To discard the Old Testament was tantamount to discarding the person of Jesus, an act that would have risked the entire faith; therefore, the Church Fathers retained the normativeness of the old scriptures by appealing to "allegory" and "typology."

Clement[37] (155–215) and Origen (185–254), who were among the school of Alexandria's theologians and philosophers, advocated this allegorical recourse, which, later on, was adopted by other Fathers like Ambrose and Augustine. Origen saw many difficulties with the literal textual sense of the scriptures[38]: "Now the reason those we have just mentioned have a false understanding of these matters is quite simply that they understand scripture not according to their spiritual meaning but according to the sound of the letter."[39] According to contemporary Biblical authority Raymond E. Brown,

> Many of the Church Fathers, e.g., Origen, thought that the literal sense was what the words said independently of the author's intent. Thus were Christ spoken of as "the lion of Judah," the literal sense for these Fathers would be that he was an animal. That is why some of them rejected the literal sense of Scripture.[40]

Origen argued that "all [scripture] has a spiritual meaning but not all a bodily meaning"[41] and observed that certain Biblical passages do not make sense at all unless understood allegorically: "Now what man of intelligence will believe that the first, second, and third day, and evening and the morning existed without the sun, moon, and stars?" Therefore, he interpreted them thoroughly and allegorically. Modern Biblical authorities such as Charles Bigg,[42] Harry A. Wolfson,[43] and Jean Danielou argue that Origen derived this method of interpretation from Philo of Alexandria, the Latinized, Platonist Jewish exegete. Bigg observes that "his rules of procedure, his playing with words and numbers and proper names, his boundless extravagance are learned, not from the New Testament, but through Philo from the puerile rabbinical schools."[44] Saint Augustine of Hippo, in the name of having sound principles for interpretation, also allegorized extensively.

From 600 to 1200, allegory had a real hold upon the minds of Christian theologians. Emil Brunner observes "the rank growth of the allegorical method of Biblical exposition made it impossible to maintain the Bible text as normative, as compared with the ecclesiastical development of doctrine." By means of allegorical exposition, the Scholastics, says Brunner, "prove, with the help of Scripture, all that they wish to prove."[45] The outcome was, as John Bright puts it,

> a wholesale and uncontrolled allegorizing of Scripture, specifically the Old Testament. This did not confine itself to difficult or morally offensive passages, or to passages that tell of something that seems unnatural or improbable, or to places where Scripture contradicts, or seem to contradict, other Scripture; it extended itself almost everywhere. Scarcely a text but yielded hidden and unsuspected riches to the interpreter's ingenuity.[46]

This process of giving the Biblical text allegorical meanings led, as Bright observes, to "the exotic jungle of fanciful interpretation."[47] That was why

Christian reformers like Martin Luther and John Calvin vehemently opposed the allegorical interpretation of the Bible. Calvin described such interpretations as an invention of the Devil, something "puerile" and "far-fetched" meant to undermine the authority of scripture.[48] These reformers emphasized the text's literal meanings. By emphasizing its plain historico-philological sense, Luther and Calvin emphasized the scripture's authority and dispensed with the Catholic "Traditions" with their accepted jungle of mystical meanings.[49]

ALLEGORICAL INTERPRETATIONS AND SOME MUSLIM SECTS

This third-century Neoplatonist ideology of allegorizing the text crept into Muslim circles by the end of the first century AH, mainly due to the efforts of people like Jaᶜad ibn Darham,[50] Jaham ibn Safwan, and Wasil ibn ᶜAta', a renegade student of Hasan al-Basri. Wasil is thought to be the founder of the Muslim speculative theology embodied by the Muᶜtazilites.

The debate about God's absolute transcendence and protecting *tawḥīd* (divine unity) from the dangers of partnership, anthropomorphism, and corporealism were the leading forces in the direction of metaphorical interpretations (*ta'wīl*). Jaham ibn Safwan treated the issue of divine attributes and names at length. He met Jaᶜad at Kufa and followed his theology. Like Jaᶜad, he emphasized God's absolute transcendence by refuting all possibilities of anthropomorphism, and metaphorically interpreted all Qur'anic verses (*ta'wīl*) that could lead to any doubt of anthropomorphism regarding God. Al-Ashᶜari reports that Jaham even denied such Qur'anic terms as *shay'*, that God is "a thing (*shay'*), because that is similarity with other things."[51]

Owing to the later influence of Jaᶜad and Jaham's theological positions on the Muᶜtazilites and others, Ibrahim Madkur calls them the title of "the founders of philosophical theology in Islam."[52] Morris S. Seale describes Jaham as the real founder of Muᶜtazilites instead of Wasil ibn ᶜAta'.[53] The recognized founder of the Muᶜtazilites, Wasil ibn ᶜAta', on the other hand, was a contemporary of Jaᶜad and Jaham. Jaham's theology, argues Hamilton A. R. Gibb,

> left distinct traces on that of the Muᶜatazilah; the doctrine of the created Kur'an which was later to become a fundamental Muᶜatazilah thesis was probably formulated by Djahm and in the doctrine of the divine attributes there are coincidences on both sides which cannot be accidental. On the other hand, there are many serious differences which are probably practical and political in their nature. Djahm

> professed in the most extreme form the doctrine of predestination (*djabr*). All the actions of man are involuntary. Wasil maintained the opposite thesis of free will.[54]

Muᶜtazilites influenced by *ta'wīl* tried to allegorize many of the Qur'anic verses and prophetic traditions. Their philosophy was strongly influenced by Neoplatonist Aristotelianism. They attempted to provide a philosophical basis for their speculative theology by metaphorically interpreting the Qur'anic text whenever it was considered to be in conflict with intellect. Although some of the later Muᶜtazilite scholars declared that the conclusions of their philosophy and the Shariᶜah were essentially harmonious, they nevertheless placed reason above revelation and especially above the prophetic insights and traditions. Consequently, they denied a large number of prophetic reports.

The Muᶜtazilites also utilized Greek logic and rationalism to support the Islamic revelation and dogmas and to convince non-Muslims of their vitality. Later, however, they gave priority to reason (*ᶜaql*) over revelation (*waḥy*), as Zahdi Jar Allah observes.[55] While the Qur'an, argues Andrew Rippin,

> had its place in the discussions, it was not so much a source, when used by Muᶜtazila, as a testimony to the veracity of the claims which they were making. The basic assumption of the Greek philosophical system (as understood and transmitted through Christian scholars) was the fundamental element underlying the whole position; it was argued that reason, and not only traditional sources, could be used as a source of reliable knowledge for human beings.[56]

This view of the role of reason, Rippin further argues, "is significant in terms of the ultimate fate of the Muᶜtazilite, for it implied that the legal scholars of Islam had, in fact, no particular claim to sole possession of the right interpretation of all Muslim dogma."[57] In addition, the Muᶜtazilites became militant once they were given political authority by such Abbasid caliphs as Harun al-Rashid and his son ᶜAbd Allah al-Ma'mun. Frederick M. Denny observes that the Muᶜtazilites,

> far from being liberal intellectuals who wanted to accommodate the world to a vision of rationality and cooperation, were proponents of a strict and militant Islam which they sought to impose uniformly on their wayward coreligionists and to spread to the non-Muslims by means of propaganda.[58]

Finally, the Muᶜtazilites metaphorically interpreted all of the Qur'anic verses that ascribed to God a face, hands, eyes, etc., and imposed these interpretations on others.[59] Despite "their several disagreements on points of doctrinal details," observes Netton, "most of the Muᶜtazilites were agreed on a non-literal mode of interpretation of much of the anthropomorphic data about God in the Qur'an." They interpreted "face" in "every thing will perish except the face of thy Lord" (28:88) to mean the being of God Himself. His hand was interpreted as "favor or bounty,"[60] the eye as "knowledge," the establishment upon the Throne (*istiwā'*) as "dominance," the descent in the later part of the night as the closeness of His "Mercy."[61] W. Montgomery Watt observes that the Muᶜtazilites dealt

> with the anthropomorphisms by the method of *ta'wīl* or "metaphorical interpretation." More precisely this meant that they claimed they were justified in interpreting single words in the Qur'anic text according to a secondary or metaphorical meaning found elsewhere in the Qur'an or in pre-Islamic poetry. Thus, in the phrase (38:75) about God "creating with his hands" they said that hands meant "grace" (*niᶜmah*), and justified this by a usage roughly parallel to our colloquial phrase "I'll give you a hand." Similarly *wajh*, usually "face," was said to mean "essence." Verses which spoke of God being seen in the world to come were interpreted in the light of other verses where "see" did not mean physical sight. In some ways this method of interpretation is artificial; but at least it keeps thinkers at the "grass roots" of religious experience and away from an abstract academic discussion of relations between attributes and essence.[62]

They treated the prophetic hadiths much more harshly if they considered that they contradicted the logical interpretations. They rejected many of these hadiths altogether, as Georges Anawati demonstrates.[63]

Despite their great contributions to Islam's intellectual life and their being the "founders of the discipline of speculative or philosophical theology," the Muᶜtazilites departed far from the spirit of Islamic revelation and hence from the outlook of ordinary Muslims. "To insist on the bare unity of God," argues Watt, "was a tidy rational theory, but it did not do justice to the fullness of religious experience. The negative statements of Dirar and an-Nazzam are unsatisfactory to the ordinary worshipper."[64] The Muᶜtazilites reduced the vivid and living God of Muhammad, as Duncan B. MacDonald argues, to "a spirit, and a spirit, too, of the vaguest kind."[65] The Muᶜtazilites, opines Fazlur Rahman, "denuded God of all content and rendered Him unsatisfactory for religious consciousness."[66] They, observes Netton, "made God more unknowable rather than less, and dug a wider gulf between man and his Creator. A dry hermeneutic intellectualism restricted the former's mental image of his Deity."[67] Such a concept, observes Watt, "leads to an abstract, bare and featureless conception of God, which robs the religious consciousness of much that is precious to it."[68] Or, in the words of Gibb, the Muᶜtazilites turned the Islamic concept of God into "a vast old monument, beneath which the element of personal religious experience seemed to be crushed out of existence. Fortunately for Islam, it was not to be so."[69]

ALLEGORICAL INTERPRETATIONS AND MUSLIM PHILOSOPHERS

On the other hand, the philosophers and later the Ismaᶜilites negated, in the name of God's unity and transcendence, His attributes by allegorizing the Qur'anic and prophetic texts. Philosophers like al-Farabi,[70] Ibn Sina,[71] and Ibn Rushd[72] simply stripped God of all the possible attributes ascribed to Him by the Qur'an, using the same allegorical method of interpretation.[73]

The Ismaᶜilites[74] followed the philosophers in this regard and in ascribing all of the divine attributes to the First Intelligence on the Neoplatonist model.[75] This First Intelligence, rather than God Himself, seemed to be the true Deity, because the God of the Ismaᶜilites and the philosophers was the bare reality and the absolute unknowable One. This God seemed to need the First Intelligence to create, sustain, protect, and love. In an effort to exalt God beyond all possible limitations and needs, the philosophers finally bound Him too tightly with their theory of emanation and hence with several limitations. Netton differentiates between the Muᶜtazilites' deity and that of the Neoplatonists' deity of the philosophers:

> The transcendent Deity of the Muᶜtazilites, whose several Qur'anic attributes were metamorphosed by allegory, was not bound up with ideas of emanation, nor with hypostases such as the Universal Intellect [*al-ᶜAql al-Kulli*] and the Universal Soul [*al-Nafs al-Kulliyya*]. But the unknowable God of medieval Neoplatonic Islam was. The end result was the development of a transcendental theology in Islam, with the Ismaᶜili sect as its political and spiritual apotheosis, which was far more complex than anything of which the Muᶜtazila could have dreamed.[76]

ALLEGORICAL INTERPRETATIONS AND THE *SALAF*

The Mu'tazilites' somewhat abstract concept of God and a thorough allegorization of the Qur'anic and prophetic texts, along with the Hashwites' anthropomorphic concept of God and absolute literalization of the Islamic texts, caused the early Orthodox scholars (*Salaf*) to devise their formula of *bilā kayf*, in other words, understanding the Islamic texts and their words as they are without asking the "how" of things. These early scholars were often called the People of Tradition (*Ahl al-Ḥadīth*) or *Salaf.* Among them were the imams Abu Hanifah,[77] Malik, Shafiʿi, and Ahmad ibn Hanbal,[78] who followed the Qur'anic verses and the prophetic hadiths as they occurred, without applying much reason either to criticize or to expand upon them, and accepted the Qur'anic statements as they stood. These conservatives, observes Majid Fakhry, "tended to repudiate the use of any deductive method"[79] and argued that ambiguous verses must be understood in the light of the clear verses. At the same time, they used and maintained the same phrases that have been implied by the Qur'an regarding God, such as *wajh Allāh* (God's face) without looking further into the "how" or "what" of them. That is what they meant by *bilā kayfa walā tashbīh* (without inquiring how and without anthropomorphism or comparison). Edward Sell observes:

> The four leading theologians, Abu Hanifa, ash-Shafiʿi, Ahmad ibn Hanbal and Malik ibn Anas, taught that discussions on such subjects were unlawful. They believed in the Qur'an and in the traditions regarding Muhammad, and accepted these without disputing on such abstract questions.[80]

Mainstream scholarship confronted this allegorization of the Islamic texts. Ahmad ibn Hanbal almost paid with his life to oppose this liberalization of the Islamic texts by some Muʿtazilites. He was severely tortured, imprisoned, and humiliated by Caliph al-Ma'mun in 202 AH during the trial over whether the Qur'an was created or eternal (and hence not created), known as the *khalq al-Qur'ān* trial. In addition, he was accused of anthropomorphism, literalism, and corporealism. On the contrary, Ibn Hanbal did not take an absolute literal approach to the Qur'anic and prophetic texts. In fact, his anti- anthropomorphism can be seen in his strong opposition to any anthropomorphic description of the Qur'anic phrases that apparently describe Allah in human terms, such as a person with hands or a face.

Al-Shahrastani reported that Ibn Hanbal said:

> Whoever moved his hand while reading the Qur'an, "I created with My hands" (38:75), ought to have his hand cut off; and whosoever stretched forth his finger in repeating the saying of Muhammad, "The heart of the believer is between two fingers of the Merciful," deserves to have his fingers torn out.[81]

Watt observes:

> There were naive anthropomorphists among the Traditionists, but he [Ibn Hanbal] opposed these as vigorously as he opposed the Muʿtazilites; he insisted that the anthropomorphic expressions of the Qur'an are to be understood "with-

> out stating the precise manner of their existence" (*bilā kayf*, literally "without how"). The strength of Ibn-Hanbal's feelings on the matter (related to preservation of the pristine letter as well as the spirit of the Islamic texts and his vehement opposition to metaphorical interpretations) may be gauged by the fact that he broke off relations with a follower who attempted to refute the Muʿtazilites by their own methods of argument.[82]

This helps show that "the Hanbalites' position was based on an awareness of the limitations of reason in this sphere, coupled with an understanding of the need to retain the concrete and 'poetical' language of the Qur'an and the Traditions."[83] In the words of Karen Armstrong, Ibn Hanbal was not anthropomorphist, but rather was "stressing the essential ineffability of the divine, which lay beyond the reach of all logic and conceptual analysis."[84]

One of the main reasons why allegorical and metaphorical interpretations (*ta'wīl*) were avoided was, as al-Shahrastani explains, that an interpretation is "an opinion ... for we may sometimes interpret the verse in a way not intended by God, and thus we would fall into perversity."[85] Following the lead of the early *Salaf*, Ibn Taymiyyah and Ibn al-Qayyam opposed metaphorical interpretations, emphasized the texts' apparent and literal sense, and tried to preserve the religious texts right through to the letter. They maintained that ambiguous verses were to be understood in the light of Arabic's fixed grammatical rules. Phrases like the "face of God" or the "hands of God" should be understood in accordance with their common, daily linguistic usages. So, the "face of God" meant His face as we understand the meaning of "face" in our daily usage, without giving it a metaphorical interpretation. Such an understanding, in their opinion, does not imply any comparison, corporeality, or anthropomorphism, for the level of His attributes is absolute, whereas in the creation this level is relative. God has already explained that no one and no thing is like unto Him, but He, at the same time, is all-hearing and all-seeing. Thus accepting that He has attributes that are also possessed by human beings does not make Him similar to them. Likewise, accepting that He has attributes like a hand and a face would not be anthropomorphic, for they are also different from human hands and faces.[86] Therefore, when we say, "God has a face or hands," it must be qualified with the qualifier "not like our face or hands" and without how. Ibn Qudamah,[87] Ibn Taymiyyah,[88] and many other traditional scholars hold this opinion.

قال الإمام الخطابي: مذهب السلف في الصفات إثباتها وإجراؤها على ظاهرها ونفي الكيفية والتشبيه عنها.[89]

Imam al-Khattabi said, "Regarding [Divine] Attributes the *Salaf's* position is to affirm them by taking them on their face value while rejecting any form/shape or likeness."

According to Ibn Taymiyyah, any non-literal meaning given to a Qur'anic phrase was an alteration (*taḥrīf*); therefore, he opposed *ta'wīl* (metaphorical interpretation).

I have tried to make the point that the *Salaf 's* insistence upon literalism should be understood in relation to their cultural milieu and its thorny challenges. It was not an ideal solution, and many classical scholars did not follow it for a variety of reasons. At the same time, this selective insistence upon some meanings at the expense of others had become a necessity for some jurists trying to counteract the dangers of liberalizing the meaning of the Qur'anic and hadith texts. The debate was much more relevant to the issues of *ʿaqīdah*, the public good (*maṣlaḥah mursalah*) and long-term consequences (*maʿālāt*) rather than such simple *fiqhī* issues as sighting the moon. The conservative jurists wanted to ascertain that the Shariʿah's pristine, simplistic, and easy-to-implement formulation was not compromised, that it did not become a prerogative of an educated élite, but rather an easily accessible commodity to all Muslims, even the least educated. Moreover, this debate over literalism and metaphorical interpretation cannot be viewed in a vacuum. Their juristic edicts must be understood in the light of their cultural milieu, historical challenges, and personal *fiqhī* dispositions. The norms and customs (*ʿurf*) of their times seem to have played a major role in how these texts were interpreted, especially in their understanding of the unity of horizons, prayer times, and the qiblah directions.

Imam al-Qarrafi, after giving examples of many custom-based rulings, observes:

وَعَلَى هَذَا الْقَانُونَ تُرَاعَى الْفَتَاوَى عَلَى طُولِ الأَيَّامِ فَمَهْمَا تَجَدَّدَ فِي الْعُرْفِ اعْتَبِرْهُ وَمَهْمَا سَقَطَ أَسْقِطْهُ وَلا تَجْمُدْ عَلَى الْمَسْطُورِ فِي الْكُتُبِ طُولَ عُمْرِكَ بَلْ إِذَا جَاءَكَ رَجُلٌ مِنْ غَيْرِ أَهْلِ إِقْلِيمِكَ يَسْتَفْتِيكَ لا تَجْرِهِ عَلَى عُرْفِ بَلَدِكَ وَاسْأَلْهُ عَنْ عُرْفِ بَلَدِهِ وَاجْرِهِ عَلَيْهِ وَأَفْتِهِ بِهِ دُونَ عُرْفِ بَلَدِكَ وَالْمُقَرَّرِ فِي كُتُبِكَ فَهَذَا هُوَ الْحَقُّ الْوَاضِحُ وَالْجُمُودُ عَلَى الْمَنْقُولاتِ أَبَدًا ضَلالٌ فِي الدِّينِ وَجَهْلٌ بِمَقَاصِدِ عُلَمَاءِ الْمُسْلِمِينَ وَالسَّلَفِ الْمَاضِينَ.[90]

Based upon this rule, fatwas [religious edicts] must be reviewed with the passage of time. The changes in *ʿurf* [customs] should be reflected in the fatwas [the revised customs should be incorporated and the discarded customs should

> be dropped]. Do not always adhere to the fatwas written in the books [throughout your life]. You should not answer a questioner according to your customs if the questioner belongs to a different province. You should ask him about the norms of his area and give a fatwa not based upon your customs and what is decided in the [*fiqhī*] books but according to the questioner's customs. This is exactly the right way. It is an act of clear misguidance in religion to continue adhering to [the fatwas] transmitted in the books, for it reflects ignorance about the real intent of the Muslim scholarship and also the intent of the early ancestors (*Salaf*).

It seems evident that the earlier jurists' fatwas on calculations as regards determining the qiblah and prayer times, as well as their insistence upon the unity of horizons, were based upon their *ʿurfī* understanding of the prophetic hadiths. Later jurists could not have deduced altogether different rulings from the same hadiths if these hadiths had been categorical in their previously understood meanings. The fact that the earlier understandings were modified and, at times, completely changed leads us to the logical conclusion that the previous interpretations were based upon the customs and realities existing at that time.

Contemporary Islamic scholarship can say the same thing about the later juristic discourse on moon sighting. We can deduce that classical scholarship rejected astronomical calculations for a variety of reasons mostly connected with their cultural milieu and their understanding of the particular prophetic hadiths. We have the equal right to review their particular *ʿurfī* interpretations in the light of our own cultural realities. In this way, we can make Islamic fiqh relevant to our times, just as al-Qarrafi and others tried to do in their particular eras. Al-Qarrafi contradicted an established *sharʿī sabab* (a legal cause established by the Maliki, Hanbali, and Hanafi schools), namely, that the unity of horizons requires all Muslims to begin fasting Ramadan and established a new legal cause, namely, the variety of horizons. Likewise, we can review this established legal cause of al-Qarrafi and other classical jurists (naked-eye sighting) and replace it with a more authentic and precise method, that of astronomical calculations. This idea will not be acceptable to everyone in the beginning. Nevertheless, with the passage of time and the appropriate awareness it will become the norm, just as it did for determining the qiblah and the daily prayer schedule.

CHAPTER 2

ANALYSIS

In view of this discussion, let us analyze the main arguments put forward for actual moonsighting.

THE QUR'AN NEVER REQUIRED NAKED-EYE SIGHTING OF THE MOON

The Qur'an is the normative authority, yet nowhere in it does Allah unequivocally require that the month of Ramadan be determined by naked-eye sighting. As he Qur'an is revealed in a clear, plain, and very straightforward Arabic, the linguistic meanings of most of its terms are obvious. The two consecutive verses in *Surat al-Baqarah* about the month of Ramadan and the act of fasting are so clear that their basic meaning and their true essence cannot be misunderstood.

(شَهْرُ رَمَضَانَ الَّذِي أُنْزِلَ فِيهِ الْقُرْآنُ هُدًى لِلنَّاسِ وَبَيِّنَاتٍ مِنَ الْهُدَى وَالْفُرْقَانِ فَمَنْ شَهِدَ مِنْكُمُ الشَّهْرَ فَلْيَصُمْهُ وَمَنْ كَانَ مَرِيضًا أَوْ عَلَى سَفَرٍ فَعِدَّةٌ مِنْ أَيَّامٍ أُخَرَ يُرِيدُ اللَّهُ بِكُمُ الْيُسْرَ وَلا يُرِيدُ بِكُمُ الْعُسْرَ وَلِتُكْمِلُوا الْعِدَّةَ وَلِتُكَبِّرُوا اللَّهَ عَلَى مَا هَدَاكُمْ وَلَعَلَّكُمْ تَشْكُرُونَ.) (البقرة: 185-186)

> Ramadan is the [month] in which was sent down the Qur'an, as a guide to humankind, also Clear [Signs] for guidance and judgment [between right and wrong]. *So whosoever witnesses the month among you should fast in it [spend it in fasting]*, but if anyone is ill, or on a journey, the prescribed period [should be made up] by days later. *Allah intends ease for you; He does not want to put you in difficulties.* [He wants you] to complete the prescribed period, and to glorify Him in that He has guided you; and perchance you shall be grateful. (2:185)

The phrase "So whosoever witnesses the month among you should fast in it" requires witnessing as a prelude and prerequisite for fasting Ramadan. Qur'anic exegetes and grammarians have reached a consensus that *fa* in *fal yaṣumhu* is causative; meaning that the clause "then let him fast" is causative and not descriptive. For instance, Shihab al-Din al-Alusi states:

ولذا ذهب أكثر النحويين إلى أن الشهر مفعول به فالفاء للسببية أو للتعقيب لا للتفصيل.[1]

That is why most of the philologists maintain that the month is the object. Therefore, the *fa* is causative or appellative and not descriptive.

It is clear that the Qur'an establishes "witnessing the month" as the sole cause (*sabab*) of fasting Ramadan. Allah has not left it to us to determine this process of witnessing, for the Qur'an itself ascertains and establishes its meanings.

The linguistic meanings of *shahida* are "presence, knowledge, and announcement (informing others)." In his *Maqāyīs al-Lughah,* Ahmad ibn Faris states:

شهد الشين والهاء والدال أصلٌ يدلُّ على حضور وعلم وإعلام، لا يخرُج شيءٌ من فروعه عن الذي ذكرناه.[2]

The original meanings of *shahida* are confined only to three: Presence, knowledge, and announcement. None of the word's derivatives go beyond these three meanings, as we have mentioned.

Linguistically, this phrase can have only three meanings:

1: Whoever was present in the month of Ramadan should fast [it].
2: Whoever had knowledge of the month of Ramadan should fast [it].
3: Whoever received knowledge of the month of Ramadan should fast [it].

In no way can it be translated as "Whoever physically sights the moon of the month of Ramadan, let him/her fast it," for this would contradict all of Arabic's established grammatical rules. Thus, all Qur'anic exegetes, without exception have translated and understood this phrase's meaning as "Whosver was present in (or knew of) the month of Ramadan, let him/her fast the month."

THE QUR'ANIC USE OF *SHAHIDA*

The Qur'an has used *shahida* in all the above-mentioned three meanings without resorting to any requirement for actual physical sighting. For instance:

(شَهِدَ اللَّهُ أَنَّهُ لا إِلَهَ إِلا هُوَ وَالْمَلائِكَةُ وَأُولُو الْعِلْمِ قَائِمًا بِالْقِسْطِ لا إِلَهَ إِلا هُوَ الْعَزِيزُ الْحَكِيمُ.)
(آل عمران: 18)

There is no god but He: that is the witness of Allah, His angels, and those endued with knowledge, standing firm on justice. There is no god but He, the Exalted in Power, the Wise. (3:18)

Neither Allah nor the angels witness with actual eyes. Al-Razi explains that the witnessing of Allah and the angels in this verse can have only two meanings: "the news that is bolstered with proof of knowledge" or "witnessing means the demonstration and exposition" of Allah's unity.

أن تجعل الشهادة عبارة عن الإخبار المقرون بالعلم. فالمفهوم الإظهار والبيان.[3]

The witnessing of Allah means that He "explains or knows."

Al-Suyuti explains the same implications of the divine act of witnessing by using the following words:

شَهِدَ اللَّهُ بيَّن لخلقه بالدلائل والآيات.[4]

Shahida here means that Allah had explained to His creatures by the signs and arguments [that He is One].

The Qur'an uses the same word to denote such human faculties as hearing and seeing. However, these human faculties do not possess actual physical eyes in order to witness things.

(حَتَّى إِذَا مَا جَاءُوهَا شَهِدَ عَلَيْهِمْ سَمْعُهُمْ وَأَبْصَارُهُمْ وَجُلُودُهُمْ بِمَا كَانُوا يَعْمَلُونَ وَقَالُوا لِجُلُودِهِمْ لِمَ شَهِدْتُمْ عَلَيْنَا قَالُوا أَنْطَقَنَا اللَّهُ الَّذِي أَنطَقَ كُلَّ شَيْءٍ وَهُوَ خَلَقَكُمْ أَوَّلَ مَرَّةٍ وَإِلَيْهِ تُرْجَعُونَ.)
(فصلت: 20-21)

At length, when they reach the [Fire], their hearing, their sight, and their skins will bear witness against them, as to [all] their deeds. They will say to their skins: "Why do you bear witness against us?" They will say: "Allah has given us speech, He Who gives speech to everything: He created you for the first time, and unto Him were you to return." (41:20-21)

In these verses, the witness of "their hearing, their sight, and their skins" is explained by *shahida*. No one can say that these faculties will witness with their eyes. Instead, this means that they will explain or give knowledge of what the person did while in the world.

The following verse uses *shahida* for truth:

(وَلا يَمْلِكُ الَّذِينَ يَدْعُونَ مِنْ دُونِهِ الشَّفَاعَةَ إِلا مَنْ شَهِدَ بِالْحَقِّ وَهُمْ يَعْلَمُونَ.) (الزخرف: 86)

And those whom they invoke besides Allah have no power of intercession; only he who bears witness to the Truth, and they know [him]. (43:86)

Here again, the witness of truth cannot be explained with the naked eye. If it could, then truth would have to be a physical entity. Since this is not the case, it means to stand by the truth or acknowledge it wholeheartedly.

In light of the clear Qur'anic and linguistic meanings, Qur'anic exegetes interpret 2:185 to mean: "Whoever was present in the month of Ramadan and was not traveling or sick should fast it."

Al-Qurtubi reports that ʿAli, Ibn ʿAbbas, ʿA'ishah, and other well-known Companions maintained that *shahida* means "be present" in the month of Ramadan:

فقال عليّ ابن أبي طالب وابن عباس وسُوَيد بن غَفَلَة وعائشة أربعة من الصحابة وأبو مِجْلَز لاحق بن حُميد وعَبيدة السَّلْمانِيّ: من شهد أي من حضر دخول الشهر وكان مقيماً في أوله في بلده وأهله فليكمل صيامه...، ومن أدركه حاضراً فليصمه.[5]

ʿAli ibn Abi Talib, Ibn ʿAbbas, Suwayd ibn Ghafalah, and ʿA'ishah, four of the Companions, and Successors such as Abu Mijlaz, Lahiq ibn Humayd, and ʿUbaydah al-Salmani have said, "*Shahida* means whosoever was present when the month started and was resident in his city and among his family, let him complete his fasting...Whosoever was present in the month of Ramadan, let him fast."

Ibn Kathir says:

(فَمَن شَهِدَ مِنكُمُ الشَّهْرَ فَلْيَصُمْهُ) فأثبت الله صيامه على المقيم الصحيح، ورخص فيه للمريض والمسافر.[6]

By this verse, Allah required the resident and healthy to observe fasting, while giving a concession to the sick and traveling persons.

Al-Suyuti says that *shahida* means being "present."

(فَمَن شَهِدَ) حضر.[7]

Al-Nasafi gives the same interpretation:

(فَمَن شَهِدَ مِنكُمُ الشَّهْرَ فَلْيَصُمْهُ) فمن كان شاهداً أي حاضراً مقيماً غير مسافر في الشهر فليصم فيه ولا يفطر.[8]

Al-Razi gives the same meanings as well and states that *al-shuhūd* means "being present."

(شَهِدَ) أي حضر والشهود الحضور.[9]

Qur'anic exegetes provide no second opinion about this meaning. In other words, they have reached a consensus that witnessing the month means nothing more than being present in the month الشهود / الحضور.

The same meanings of "presence" are also conveyed by the context in which this phrase occurs, for immediately after it the Qur'an states: "Whoever is sick or traveling should make up for the missed days." Imam al-Alusi observes:

ولا يحسن أن يقال من علم الهلال فليصم ومن كان مريضاً أو على سفر فليقض لدخول القسم الثاني في الأول والعاطف التفصيلي يقتضي المغايرة بينهما...ولذا ذهب أكثر النحويين إلى أن الشهر مفعول به ـــ فالفاء ـــ للسببية أو للتعقيب لا للتفصيل.[10]

Linguistically, it is not appropriate to say, "whosoever knew [saw] the moon, let him fast, and whosoever is sick or traveling, let him make up for when the second category enters the first category" (meaning when the sick or traveling person knows the new moon). The detailed conjunction "and" demands difference and variety… that is why the majority of the grammarians view *al-shahr* as the object and maintain that the word *fa* is causative, meaning "as a result" or "consequently" [the person should fast], and it is not descriptive.

Al-Alusi shows that in linguistic terms, the phrase cannot violate the two established meanings. It has to mean presence in person or from knowledge or an announcement. These meanings do not give a sense of actual naked-eye sighting regardless of whether we consider *al-shahr* as *maf'ūl fīhi* or *maf'ūl bihi.*

(شَهِدَ) من الشهود والتركيب يدل على الحضور إما ذاتاً أو علماً، وقد قيل: بكل منهما هنا.[11]

He concludes:

فمن حضر في الشهر ولم يكن مسافراً فليصم فيه أو من علم هلال الشهر وتيقن به فليصم.[12]

Whoever was resident during the month of [Ramadan] and not traveling or knew the start of the new moon of the month with certainty should fast.

Al-Razi also states that in either case, the meaning will be "presence," not "witnessing the new moon with the eye."

ثم ههنا قولان: أحدهما: أن مفعول شهد محذوف لأن المعنى: فمن شهد منكم البلد أو بيته بمعنى لم يكن مسافراً وقوله: (الشَّهْرُ) انتصابه على الظرف وكذلك الهاء في قوله: (فَلْيَصُمْهُ). والقول الثاني: مفعول (شَهِدَ) هو (الشَّهْرُ) والتقدير: من شاهد الشهر بعقله ومعرفته فليصمه وهو كما يقال: شهدت عصر فلان، وأدركت زمان فلان.[13]

There are two views [regarding this phrase]. First is that the object of witnessing is omitted because the meaning is that "whoever witnessed his residence [city or house] and was not traveling [should observe fasting]". ... The second view is that the object of witnessing is the month, meaning "whoever witnessed [knew] the month from his knowledge and intellect should fast it." It is just like saying phrases such as "I witnessed such and such person's era..."

It is a common Arab practice to say "I witnessed the Friday prayers" or "I witnessed the hajj." That does not mean that either event is something physical and the person saw it with his/her own eyes; rather, it means that one was present during the Friday prayers or the hajj of a particular year.

Al-Razi concludes that witnessing the month can be done with two methods: actual sighting or hearing about it.

أن شهود الشهر بماذا يحصل؟ فنقول: إما بالرؤية وإما بالسماع.[14]

How is *shuhūd al-shahr* ("witnessing the month") accomplished? We say that it is achieved by either physical sighting or hearing.

SOME WRONG INTERPRETATIONS

The above analysis shows the true meaning of *shahida* and its essence within the Qur'anic context. The verse, which is "decisive in its transmission and unequivocal in its meaning," is not limited to naked-eye sighting, although some contemporary Muslims claim that such sighting is the only acceptable etymology of this word. For instance, Zaheer Uddin, Coordinator for the Hilal Sighting Committee of North America, along with the approval of "the great Ulema of the Shura of the Hilal Sighting Committee of North America," argues: "The first evidence we rely on is based on the Qur'an." After quoting 2:185, he observes:

The critical term in this verse is "*faman shahida minkum as-shahra,*" which means to the effect "those from you who have witnessed the month." This verse has also been suggested to mean "those who are present at their home." The meaning is still pointed towards presence and witnessing the month.

He further contends:

The root meaning of the critical term "*shahida*" refers to a **witnessing which takes part in a physical form, as a form of proof.** Scholars of Arabic grammar agree on this by unanimous consensus. Ibn Abbas (r), who was the first man to collect Arabic words and elaborate its meaning, also clearly holds this view. Khalil, who used Ibn Abbas's collection to formulate the world's first dictionary in any language (this of course being in Arabic), published this fact in his works, "*Ayn*". Scholars who study grammar consider this work to be the most authoritative in understanding the depths of meaning of Arabic words. "*Shahada*" cannot happen in a simulated environment. It has to occur in a physical form as a form of proof. Of course, the result of "*shahida*", i.e., the witnessing can be recorded, and this should not be misunderstood with the witnessing itself."[15]

Unfortunately, sometimes we get so carried away by our positions that we make sweeping statements that are both unwarranted and unsubstantiated by the actual texts. I have yet to see the above statement attributed to Ibn ʿAbbas in any authentic Islamic source, for it does not exist in the way it has been presented.[16] There is no reference to *shahida* as meaning "a witnessing which takes part in a physical form, as a form of proof" in Khalil's compilation called *Al-ʿAyn* either. It is always good to quote the original texts to avoid any mishap or misrepresentation, for they can guide the reader to the originally intended meanings.

Abu Bakr al-Jassas and ibn al-ʿArabi's Interpretations

Both Abu Bakr al-Jassas and Ibn al-ʿArabi have interpreted *shahida* in the light of authentic prophetic reports to mean "sighting the new moon." I reiterate here that these classical jurists strongly opposed calculations due to the historical facts prevalent during their time and their desire to protect the ordinary Muslims' *ʿaqīdah* (belief) from dereliction:

As for the second reason, it is not permissible [wrote Ibn al-ʿArabi] to rely on astronomers and mathematicians, not because their findings are not true but because people's beliefs must be protected from an association with celestial motions and future occurrences of conjunctions and separations. Indeed, that is a vast ocean should people be pulled into it.[17]

Their intended goal was to attain certainty in those matters connected with our acts of devotion, as Abu Bakr al-Jassas has rightly explained.

حَتَّى يَدْخُلَ فِي الْعِبَادَةِ بِيَقِينٍ ، وَيَخْرُجَ عَنْهَا بِيَقِينٍ.[18]

This is in order that we enter our [time-dependent] acts of worship with *certainty and end them with certainty.*[19]

During the early centuries of Islam, actual sighting or the completion of thirty days were the only two methods available to ascertain the month. Thus, these scholars closely adhered to these methods to attain certainty. Physical sighting was not an end in itself, but rather a method to reach the goal of required certainty. Al-Qurtubi summarized this point, as Hamza Yusuf has highlighted it:

God says, "Whoever witnesses the month, let him fast" (2:185). He means, and God knows best, "Whoever among you knows, *with a knowledge that is certain*, that the month has indeed begun must fast it." And knowledge that is certain is [based on] either a clear and widespread sound sighting or the completion of thirty days of the previous month.[20]

I also believe that the classical scholars would have reacted very differently had astronomy and calculations reached the same level of precision as that of the twenty-first century.

With due respect to the scholarship of Abu Bakr al-Jassas and Ibn al-ʿArabi, both of them tried to deduce possible Islamic rules (*aḥkām*) from the Qur'an based upon their knowledge, *fiqhī* dispositions, and culture. Their interpretations cannot possibly be granted the divine authority enjoyed by the Qur'an. In addition, neither of them ever asserted that their understanding and interpretation of *shāhid* as meaning "to see the new moon" was the only possible etymology of the word's original sense. Both of them clearly stated that *shāhid* in the Ramadan verse, in light of the prophetic reports, means "sighting the new moon." They never asserted that naked-eye sighting was the definitive original (linguistic) meaning of *shahida*.

Ibn al-ʿArabi states:

الْمَسْأَلَةُ الثَّالِثَةُ: قوله تعالى: (فَمَنْ شَهِدَ مِنْكُمْ الشَّهْرَ فَلْيَصُمْهُ): مَحْمُولٌ عَلَى الْعَادَةِ بِمُشَاهَدَةِ الشَّهْرِ، وَهِيَ رُؤْيَةُ الْهِلَالِ.[21]

The third inquiry regarding the divine statement "whoever witnesses the month, then let him fast" is that [this phrase] is usually carried [to mean] witnessing the month, which is sighting the new moon.

I was a little surprised by Hamza Yusuf's translation of this statement:

The next point is God's word in the verse, "So whoever witnesses the month, let him fast" (2:185). This is normally understood to mean, "*see it with one's eyes*" – in other words, sighting the crescent moon.[22]

Perhaps this is an unintended oversight, given his expertise in classical Arabic, for it is quite challenging to translate Ibn al-ʿArabi's rendition مَحْمُولٌ عَلَى الْعَادَةِ بِمُشَاهَدَةِ الشَّهْرِ as "is normally understood to mean 'see it with one's eyes'" in the absence of a commanding clause.

Al-Jassas equates actual sighting with "witnessing the month" of Ramadan, as we can read in the chapter describing the "Mode of Witnessing the Month" بَابُ كَيْفِيَّةِ شُهُودِ الشَّهْرِ, and not in the actual meanings of the Qur'anic verse. Like Ibn al-ʿArabi, he interprets that phrase to mean an "actual sighting," based upon the prevailing understanding of the above-mentioned verse and the relevant prophetic reports available during his time.[23] Below is his actual quote:

قَوْلُ رَسُولِ اللَّهِ صلى الله عليه وسلم: "صُومُوا لِرُؤْيَتِهِ" مُوَافِقٌ لقوله تعالى: (يَسْأَلُونَكَ عَنْ الاهِلَّةِ قُلْ هِيَ مَوَاقِيتُ لِلنَّاسِ وَالْحَجِّ) وَاتَّفَقَ الْمُسْلِمُونَ عَلَى مَعْنَى الآيَةِ وَالْخَبَرِ فِي اعْتِبَارِ رُؤْيَةِ الْهِلالِ فِي إيجَابِ صَوْمِ رَمَضَانَ ، فَدَلَّ ذَلِكَ عَلَى أَنَّ رُؤْيَةَ الْهِلالِ هِيَ شُهُودُ الشَّهْرِ.[24]

The statement of the Prophet "Fast by sighting it" is in line with the Qur'anic verse that says, "They ask you about the new moons. Say: 'They are timings for people and for hajj.'" The Muslims have agreed that the verse and the hadith mean considering the sighting of the crescent moon in requiring the fasting of Ramadan. It leads to [the fact] that sighting the new moon is witnessing the month.

This statement could be misleading if taken out of its original context, for it must be understood in relation to his overall understanding of "whoever witnesses the month should fast it." His linguistic understanding and its implications are identical with the detailed exegetical positions given above. In addition, he confines the original meanings of *shahida* to "residence, knowledge, and *taklīf*, that is, commissioning fasting with an announcement."

وقَوْله تَعَالَى: (فَمَنْ شَهِدَ مِنْكُمْ الشَّهْرَ) يَعْتَوِرُهُ مَعَانٍ، مِنْهَا: مَنْ كَانَ شَاهِدًا يَعْنِي مُقِيمًا غَيْرَ مُسَافِرٍ.[25]

The divine statement "Whoever witnesses the month should fast [it]" lends itself to a number of meanings. One of them is that "whoever witnesses means that he/she is resident and not traveling."

He clearly maintains that شُهُودُ الشَّهْرِ (witnessing the month) is the residence in presence (الاقَامَةَ فِي الْحَضَرِ) and knowing about the coming of the month:

وَقَدْ أُرِيدَ بِشُهُودِ الشَّهْرِ الْعِلْمُ بِهِ.[26]

Witnessing the month here meant knowing [arrival of] the month.

and that *shuhūd al-shahrī* means "whoever knew the month."

وَيَحْتَمِلُ قَوْلُهُ: (فَمَنْ شَهِدَ مِنْكُمْ الشَّهْرَ فَلْيَصُمْهُ) أَنْ يَكُونَ بِمَعْنَى شَاهِدِ الشَّهْرِ أَيْ عَلِمَهُ، وَيَحْتَمِلُ قَوْلُهُ: (فَمَنْ شَهِدَ مِنْكُمْ الشَّهْرَ) فَمَنْ شَهِدَهُ بِالتَّكْلِيفِ؛ لأنَّ الْمَجْنُونَ، وَمَنْ لَيْسَ مِنْ أَهْلِ التَّكْلِيفِ فِي حُكْمِ مَنْ لَيْسَ بِمَوْجُودٍ فِي انْتِفَاءِ لُزُومِ الْفَرْضِ عَنْهُ، فَأَطْلَقَ اسْمَ شُهُودِ الشَّهْرِ عَلَيْهِمْ، وَأَرَادَ بِهِ التَّكْلِيفَ.[27]

It is possible to render this divine statement as meaning, "knowing the month or knowing it while in the state of *al-taklīf* (where it is obligatory for him to fast. A traveler or a sick or insane person is not in a state of *al-taklīf*)… He [Allah] here applies the phrase "witnessing the month" to them, but in essence means the *taklīf* itself.

وَالاحْكَامُ الْمُسْتَفَادَةُ مِنْ قَوْلِهِ: (فَمَنْ شَهِدَ مِنْكُمْ الشَّهْرَ فَلْيَصُمْهُ) إِلْزَامُ صَوْمِ الشَّهْرِ مَنْ كَانَ مِنْهُمْ شَاهِدًا لَهُ، وَشُهُودُ الشَّهْرِ يَنْقَسِمُ إِلَى أَنْحَاءٍ ثَلاثَةٍ: الْعِلْمُ بِهِ، مِنْ قَوْلِهِمْ: شَاهَدْتُ كَذَا وَكَذَا؛ وَالاقَامَةُ فِي الْحَضَرِ، مِنْ قَوْلِكَ: مُقِيمٌ وَمُسَافِرٌ وَشَاهِدٌ وَغَائِبٌ؛ وَأَنْ يَكُونَ مِنْ أَهْلِ التَّكْلِيفِ عَلَى مَا بَيَّنَّا.[28]

Following is the ruling deduced from the divine statement "Whoever witnesses the month, let him fast." It requires the one who witnesses the month among them to observe fasting. The "witnessing of the month" is divided into three areas. (1) Knowledge of it [the month], for they say, "I witnessed such and such" (meaning knew such and such). (2) The residence in presence, for it is said "resident and traveling, present and absent." (3) And, as we have already explained, that the person must be of those people who are commissioned to fast.

In fact, al-Jassas would be wrong even if he insisted upon equating witnessing the month with only naked-eye sighting. Faysal Mowlawi states:

نخلص من ذلك أنّ شهود الشهر عند اللغوين وجمهور المفسرين يعني حضوره: أمّا قول الجصاص بأنّ (رؤية الهلال هي شهود الشهر) فغير مسلَّم، لأنّ رؤية الهلال تحدد بداية الشهر لا أكثر، وهي أمر يتعلّق بالهلال وبالشهر، أمّا الشهود فهو أمر يتعلّق بالإنسان المكلّف ومتى يجب عليه الصيام.[29]

From [the previous discussion] we conclude that, according to lexicograpers and the majority of Qur'anic exegetes, the phrase "wintessing the month" means "being present in the month." The statement of al-Jassas that "sighting the new

> moon is in essence witnessing the month" is unacceptable because the moon sighting does nothing but to indicate the beginning of a month. The sighting is connected with the new moon as well as the month,while the act of witnessing is connected with the person who is obliged to fast and the timings when fasting becomes obligatory upon that person.

Mowlawi further observes that the true meanings of the Qur'anic verse indicate that the sole legally binding cause for Ramadan is the coming of the month.

> وإذا كان شهود الشهر يعني حضوره، فإنّ مقتضى ذلك أنّ السبب الشرعي لوجوب صيام رمضان حسب الآية الكريمة هو (دخول الشهر)، وهو السبب الموضوعي الذي يتعلّق بالمسلمين جميعاً. أمّا الشروط المطلوبة من كل منهم حتّى يجب عليه الصيام، فهي أن يكون مقيماً غير مسافر، سليماً غير مريض، عاقلاً غير مجنون، بالغاً غير صغير، (طاهرة غير حائض إذا كانت امرأة).[30]

> If "witnessing the month" means "being present in the month" then the legal Islamic cause of Ramadan fasting obligation, according to the Qur'anic verse, is "arrival of the month [of Ramadan]. This arrival is the situational cause addressed to all the Muslims. The specific required conditions [causes] for fasting obligation include that the person be resident and not traveling, healthy and not sick, sane and not insane, adult and not a child, in the state of purity and not during her monthly cycle [in case of a female]."

Ibn al-ʿArabi maintains the same position as that of al-Jassas. They both explain that the mode of witnessing the month as leading to physical sighting, based upon arguments external to the wording of the Qur'anic phrase itself, namely, in light of the supporting evidence from the hadiths.

The Qur'anic text does not require naked-eye sighting to begin observing the Ramadan fast. In fact, it requires presence in the month, knowledge of the month, the commissioned status of a Muslim during this month, and the announcement of the month as the prelude to fasting. Actual moon sighting or *ikmāl* have always been the two established modes of achieving certainty in this regard. But they are just the means, not options or replacements. It would be disastrous to confine this particular Qur'anic command to naked-eye sighting alone, for doing so would mean that only those who actually saw the new moon would be required to fast the month of Ramadan. Those who had not seen it would be exempt, for the cause of fasting would not have been realized in their case. Moreover all Muslims, regardless of whether they were traveling, sick, chronically ill, children, or pregnant and nursing women who had actually seen the new moon would be required to fast the month of

Ramadan, because translating "witnessing" as "sighting" means that the cause (*sabab*) would have already occurred in their case. No jurist can require such a burdensome demand of those Muslims to whom Allah has granted a concession. Such legal injunctions are clear and thus beyond any other interpretation or compromise.

Consequently, after this lengthy analysis of *shahida* and all of its pursuant philological implications, I conclude that the Qur'an requires presence, knowledge of the month, and the commissioned status of a Muslim as the cause of fasting Ramadan, as opposed to actual physical sighting.

Al-Qarrafi, a Maliki, summarizes a discussion about the meanings of *shahida* in the following quotation:

وَأَمَّا قَوْله تَعَالَى (فَمَنْ شَهِدَ مِنْكُمْ الشَّهْرَ فَلْيَصُمْهُ) فَلا دَلالَةَ فِيهِ عَلَى هَذَا الْمَطْلُوبِ قَالَ أَبو عَلِيٍّ؛ لانَّ شَهِدَ لَهَا ثَلاثُ مَعَانٍ شَهِدَ بِمَعْنَى حَضَرَ وَمِنْهُ شَهِدْنَا صَلاةَ الْعِيدِ، وَشَهِدَ بَدْرًا، وَشَهِدَ بِمَعْنَى أَخْبَرَ وَمِنْهُ شَهِدَ عِنْدَ الْحَاكِمِ أَيْ أَخْبَرَهُ بِمَا يَعْلَمُهُ، وَشَهِدَ بِمَعْنَى عَلِمَ وَمِنْهُ قَوْله تَعَالَى (وَاَللَّهُ عَلَى كُلِّ شَيْءٍ شَهِيدٌ) أَيْ عَلِيمٌ وَهُوَ فِي الايَةِ بِمَعْنَى حَضَرَ قَالَ وَتَقْدِيرُ الايَةِ فَمَنْ حَضَرَ مِنْكُمْ الْمِصْرَ فِي الشَّهْرِ فَلْيَصُمْهُ أَيْ حَاضِرًا مُقِيمًا احْتِرَازًا مِنْ الْمُسَافِرِ فَإِنَّهُ لا يَلْزَمُهُ الصَّوْمُ، وَإِذَا كَانَ شَهِدَ بِمَعْنَى حَضَرَ لا بِمَعْنَى شَاهَدَ وَرَأَى لَمْ يَكُنْ فِيهِ دَلالَةٌ عَلَى اعْتِبَارِ الرُّؤْيَةِ وَلا عَلَى اعْتِبَارِ الْحِسَابِ أَيْضًا فَإِنَّ الْحُضُور فِي الشَّهْرِ أَعَمُّ مِنْ كَوْنِهِ ثَبَتَ بِالرُّؤْيَةِ أَوْ بِالْحِسَابِ.[31]

There is no proof of that [physical moon sighting or mathematical calculations] in the divine statement "whoever witnesses the month, then let him fast." Abu ʿAli stated that *shahida* consisted of three meanings: (1) "he witnessed," meaning that he was present, for it is said, "we witnessed the ʿId prayer" and [so and so] witnessed [the Battle of] Badr; (2) "he was informed," for it is said that "he bore witness in front of the judge [ruler]," when he informed the ruler about what he knew; (3) "he knew or had knowledge." The verse that says "And Allah witnesses everything" means that He possesses knowledge of everything. [The word *shāhid* in the Qur'anic verse 2:185 means "whoever was present." He [Abu ʿAli] maintained that the meaning of the verse was "whoever was present in the city [of residence] during the month [of Ramadan] then let him fast. It means whoever was present in residence and not traveling because the traveler is not required to fast. Therefore, if the meaning of the word *shahida* is to be "present" and not the act of "witnessing by the actual sighting," then it proves that there is no consideration given to the actual sighting [of the new moon] or to calculation in this verse. The presence in the month covers much more than what could be confined only to actual sighting or calculation.

Unfortunately, people sometimes try to impose their sincerely held (but mistaken) positions, and very often culturally bound dispositions, upon the

Qur'an. At times, some of them even do not let the Qur'an speak to us. We venture to inflict our reading and understanding of a given issue upon the unambiguous and self-explanatory Qur'anic texts and then, perhaps even sincerely, present it with the divine propriety. The current question is a good example of how Muslims sometimes try to compel the Qur'an to say what we think is right and should have been said. Ibn Abi Malika reminds us:

وعن ابن أبي مليكة قال: سئل أبو بكر الصديق رضي الله عنه عن تفسير حرف من القرآن فقال: أيّ سماء تُظلّني، وأيّ أرض تُقِلّني وأين أذهب وكيف أصنع إذا قلت في حرف من كتاب الله بغير ما أراد تبارك وتعالى.[32]

Ibn Abi Malika narrates that Abu Bakr was once asked about the interpretation of a word in the Qur'an. He replied, "Which heaven will cover me and which earth will carry me and where will I go and what will I do if I end up saying a word in [explaining] the Qur'an that Allah had not intended (for that Qur'anic word).

Surprisingly, the word Abu Bakr was asked to explain was not very complicated.

سئل أبو بكر الصديق رضي الله عنه عن تفسير الفاكهة والأب فقال: أيُّ سماء تُظِلني، وأيُّ أرض تُقِلُّني إذا قلت: في كتاب الله ما لا أعلم.[33]

Asked about the meanings of such simple words as *fākiha* and *al-ab*, both found in *Surat ʿAbasa*, Abu Bakr replied that he was unsure of their exact meanings in that specific context. Therefore, he was afraid and hesitant to give an interpretation that might differ from the one intended by Allah. If such was the case with him, then what about us and our finite expertise?

The Various Meanings of *al-Hilāl* in Arabic

The definition of *hilāl* as the new moon of Ramadan's first two or three nights is based upon customary or cultural meanings, not its original linguistic roots. Its linguistic meanings or etymology is not intrinsically linked with light or physical sighting; rather, it is connected with an "announcement," "raising the voice," or "the beginning" of something. The first light of the new moon, when people see it and talk about it, is also called *hilāl*. Al-Fayruzabadi informs us that a male snake, a spearhead, a small amount of water, a feeble camel, dust, a handsome young man, paved stones, and the onset of a shower are all called *hilāl* in Arabic.[34]

Al-hilāl is the new moon's shining white crescent seen at the beginning of the new month. It is said that the new moon is called *hilāl* for the first two nights, after which it is called *qamar;* that it is called *hilāl* for the first three nights only; that it is called *hilāl* until it is a quarter moon; that it is called *hilāl* until its glitter stands out brightly against the night's darkness, something that cannot happen until the seventh night; and that moon of the month's last two nights (26 and 27) is also called *al-hilāl.* Al-Fayruzabadi has this to say about the noun *al-hilāl.*

> الهِلالُ غُرَّةُ القَمَرِ أو لِلَيْلَتَيْنِ أو إلى ثلاثٍ أو إلى سبعٍ، ولِلَيْلَتَيْنِ من آخِرِ الشهرِ، سِتٍّ وعشرينَ وسبعٍ وعشرينَ، وفي غيرِ ذلك قَمَرٌ، والماءُ القليلُ، والسِنانُ، والحَيَّةُ، أو الذَّكَرُ منها، وسِلْخُها، والجَمَلُ المَهْزولُ ... والغُبارُ، ... والغلامُ الجميلُ، ...والحِجارةُ المَرْصوفَةُ... والدُّفْعَةُ من المَطَرِ ج أهِلَّةٌ وأهالِيلُ.[35]

Originally, *hilāl* is derived from the root *hallala.* Jamal al-Din ibn Manzur, the authoritative source on the Arabic lexicon, elaborates on that particular root's meanings:

> هلل: هَل السحابُ بالـــمطر وهَل الـــمطر هَلاًّ وانْهَل بالـــمطر انْهِلالاً واسْتَهَل: وهو شدَّة انصبابه. وفـــي حديث الاستسقاء: فألَّف الله السحاب وهَلَّتنا. قال ابن الأثـــير: جاء فـــي رواية لـــمسلـــم، يقال: هَل السحاب إِذا أَمطر بشدَّة، والهِلال الدفعة منه، وقـــيل: هو أَوَّل ما يصيبك منه، والـــجمع أَهِلَّة علـــى القـــياس، وأَهالـــيل نادرة. وانْهَل الـــمطر انْهِلالا: سالَ بشدَّة، واستهلَّت السماءُ فـــي أَوَّل الـــمطر، والاسم الهِلالُ. وقال غيره: هَلَّ السحاب إِذا قَطَر قَطْراً له صوْت.[36]

The cloud poured down rain…the [verb] means the pouring forth of rain… The cloud gushes forth when it rains with intensity and *al-hilāl* is the [first] burst [or the gush] from it. [It is also said that *al-hilāl* is the first gush], that is, "what you receive from that rain." Based upon this analogy, its plural form is *ahillah* and, very rarely, *ahālīl*…. Others say, "The cloud *halla*," when the rain falls in noisy drops [that rain is also called *hilāl*].

This word's root mainly has two basic meanings: to shout or to raise one's voice and begin. Ibn Manzur explains this clearly:

> انهلَّت السماءُ إِذا صبَّت، واستهلَّت إِذا ارتفع صوتُ وقعها، وكأَنَّ اسْتِهلال الصبـــيّ منه. وفـــي حديث النابغة الـــجعديّ قال: فَنـــيَّف علـــى الـــمائة وكأَنَّ فاهُ البَرَدُ الـــمُنْهَل؛ كل شيء انصبَّ فقد انْهَل، يقال: انهلَّ السماء بالـــمطر ينهلُّ انْهِلالاً وهو شدَّة انْصِبابه. قال: ويقال هلَّ السماء بالـــمطر هَلَلاً، ويقال للـــمطر هَلَلٌ وأُهْلول. والهَلَل: أَول الـــمطر. يقال: استهلَّت السماء وذلك فـــي أَول مطرها. ويقال: هو صوت وَقْعِه. واستهل الصبـــيُّ بالبُكاء: رفع صوته وصاح عند الوِلادة. وكل شيء ارتفع صوتُه فقد استهلَّ. والإِهْلال بالـــحج: رفعُ الصوت بالتَّلْبـــية. وكلُّ متكلـــم رفع

صوته أَو خفضه فقد أَهَلَّ واستهلَّ وفي الحديث: الصبيُّ إِذا وُلِد لم يُورَث ولم يَرِثْ حتى يَسْتَهِل صارخاً. وفي حديث الجَنِين: كيف نَدِي مَن لا أَكَل ولاشَرِبَ ولا اسْتَهَل.[37]

When the rain falls heavily from the sky, when someone talks loudly, when a baby cries lustily, when the pilgrim raises his voice [in declaring] *talbiyah*, all these acts are described with the same root verb *halla. Ihlāl* is simply the "raising of the voice," as all the Qur'anic exegetes, hadith interpreters, grammarians, and philologists unanimously maintain. These linguistic usages of the word *halla* and *ahlāl*, meaning "raising of the voice," coincide with the exact use of the word in many Qur'anic verses and prophetic hadiths, as Ibn Manzur elaborates.

وأَصله رَفْعُ الصَّوت. وأَهَلَّ الرجل واستهلَّ إِذا رفع صوتَه. وأَهَلَّ المُعْتَمِرُ إِذا رفع صوتَه بالتَّلْبِية، وتكرر في الحديث ذكر الإِهْلال، وهو رفعُ الصوت بالتَّلبِية. أَهَلَّ المحرِمُ بالحج يُهِلُّ إِهْلالاً إِذا لَبَّى ورفَع صوتَه. والمُهَل، بضم الميم: موضعُ الإِهْلال، وهو الميقات الذي يُحْرِمون منه، ويقع على الزمان والمصدر. الليث: المُحرِمُ يُهِل بالإِحْرام إِذا أَوجب الحُرْم على نفسه؛ تقول: أَهَل بحَجَّة أَو بعُمْرة في معنى أَحْرَم بها، وإِنما قيل للإِحرام إِهْلال لرفع المحرِم صوته بالتَّلْبية. والإِهْلال: التلبية، وأَصل الإِهْلال رفعُ الصوتِ. وكل رافِعٍ صوتَه فهو مُهِلّ.[38]

Here, Ibn Manzur shows that the origin of *hilāl* is "raising the voice." The same verb is used to define a person who raises his/her voice while declaring the *talbiyah* (chanting specific *takbīrāt* after putting on one's *iḥrām* for hajj or ʿ*umrah*). Anything that is noisy can be called *muhill.*

THE *HILĀL* IS CALLED *HILĀL* BECAUSE PEOPLE ANNOUNCE ITS SIGHTING

After a lengthy discussion of the various usages of the root word, Ibn Manzur concludes that the origin of *hilāl* is "raising the voice," not the actual sighting:

قال أَبو العباس: وسمي الهلالُ هِلالاً لأَن الناس يرفعون أَصواتهم بالإِخبار عنه.[39]

Abu al-ʿAbbas said that the *hilāl* [new moon] is named *hilāl* because the people raise their voices to inform one another about it [the new moon].

Abu Hayyan Muhammad ibn Yusuf al-Andalusi maintains the following about the word's root:

وسمي هلالاً لارتفاع الأصوات عند رؤيته من قولهم: استهل الصبي، والإهلال بالحج، وهو رفع الصوت بالتلبية، أو من رفع الصوت بالتهليل عند رؤيته.[40]

The [new moon] was called *hilāl* owing to people raising their voices on seeing it. It is derived from their expressions such as "the baby cried." The *ihlāl* for the hajj means raising the voice when declaring *al-talbiyah*. It is [also called *hilāl* because they] say out loud: "God is Great" on sighting the new moon.

The phrase وأصله رفْع الصوت (the origin of *hilāl* is "raising the voice"[41]) appears so often in the hadith literature and all known Arabic dictionaries that it is not hard to deduce that the origin of *hilāl* is that of people raising their voices and informing one another about the new month's beginning, not of the new moon's appearance. Since pre-modern people had no way of knowing the moon at the new month's beginning or the month's end and no way of passing this information on except by naked-eye sighting, sighting the new moon became correlated with the word's original linguistic meanings. As a result, the new moon of one to seven nights, as well as the last two nights of each month, were called *al-hilāl*. At this early stage it is still called *qamar*, for the Arabs used to say "*ahall al-qamar*" (the moon [*qamar*] appeared). Al-Layth even argues that the moon of the first or second night must not be referred to as *ahalla al-hilāl*, but as *ahalla al-qamar*:

قال: والهِلال غُرَّةُ القمر حين يُهِلُّه الناس في أول الشهر. تقول: أُهلَّ القَمَرُ. ولايقال أُهلَّ الهلالُ.[42]

Others, among them al-Azhari, have disagreed.

Al-Zamakhshari, the authority on Arabic as well as *tafsīr*, differentiates between the active and passive use of the word's root. In the passive sense, it means "they raised their voices on seeing the new moon," and in the active sense, it means "the new moon appeared; that is, it was seen."

وأهلّوا الهلال واستهلّوه: رفعوا أصواتهم عند رؤيته، وأهلّ الهلال واستهلّ إذا أبصر. وأهلّ الصبيّ واستهلّ إذا رفع صوته بالبكاء. وانهلّت السماء بالمطر واستهلّت وهو صوت المطر. وتهلّل السحاب بالبرق: تلألأ.[43]

I am not denying that seeing the new moon and the participation of people in sighting it are part of the meaning of *al-hilāl*. Rather, I am arguing that they are secondary meanings that became culturally popular due to pre-modern Arabs' dependence upon naked-eye sighting of the new moon to begin and end their months. The *hilāl* was called *al-hilāl* because of its original meanings: leading to an announcement and talking about something loudly. Al-Alusi makes this point concisely when explaining the verse "they ask you about the new moons."

> والأهلة جمع هلال واشتقاقه من استهل الصبي إذا بكى وصاح حين يولد ومنه أهل القوم بالحج إذا رفعوا أصواتهم بالتلبية، وسمي به القمر في ليلتين من أول الشهر، أو في ثلاث أو حتى يحجر وتحجيره أن يستدير بخط دقيق وإليه ذهب الأصمعي أو حتى يبهر ضوءه سواد الليل، وغيا ذلك بعضهم بسبع ليال وسمي بذلك لأنه حين يرى يهل الناس بذكره أو بالتكبير؛ ولهذا يقال: أهلّ واستهل ولا يقال هلّ،[44]

> *Al-ahillah* is the plural of *hilāl.* The word is derived from the crying and screaming of the newborn baby at its birth. From this root the *ihlāl* of the hajj took its name, and that happened when people began saying the *talbiyah* out loud. The moon of the first two, three… seven nights of the month was given this name also. The moon was thus named because people, when it is seen, rejoice by talking about it or by chanting the *takbīr* (*Allāhu Akbar*: God is Great). That is why it is called "*ahalla wa istahall*" and not "*halla*" (meaning "it appeared").

That is why the month is also referred to as *al-hilāl,* الْهِلالُ هُوَ الشَّهْرُ بِعَيْنِهِ "the month itself is *al-hilāl.*"[45]

The crux of the matter is that this word's original root meanings are intrinsically linked to raising the voice, informing about something publicly, and talking about something loudly. The new moon acquired its name from the fact that, metaphorically speaking, it was sighted and talked about. These culturally based metaphorical meanings became popularity owing to the pre-modern Arabs' total dependence upon the new moon's light to determine the beginning and end of their lunar months. Naked-eye sighting was the only method that achieved certainty. Gradually, the word's metaphorical meanings took over its original root meanings.

Putting the Cart before the Horse

After a detailed discussion of the subject, however, Hamza Yusuf draws a conclusion that is the exact opposite of the one maintained by the overwhelming majority of exegetical works and Arabic lexicons: "It appeared" (*halla*) or "It was seen" (*uhilla*) both entail witnessing, which means its birth is not an active event but entails, in this case, witnessing. "The moon was born" (*uhilla*) literally means, "It was seen."[46] He quotes al-Raghib al-Asfahani: "[*Ihall*] can also refer to the cry one makes upon sighting the crescent, which was later used metaphorically to refer to a baby's cry upon being born (*ihlāl al-ṣabī*)."[47]

It is pertinent to note that al-Raghib al-Asfahani is perhaps the only exegete and linguist who goes against the opinion, which has reached consensus, outlined above: he interprets "sighting of the moon" according to the original meanings of *al-hilāl* and makes the announcement or raising of one's voice part of the secondary meaning. The overwhelming majority of exegetes and linguists maintain the absolute opposite. They maintain that the original meanings of the term *al-hilāl* revolve around announcement while the actual sighting is secondary. This understanding goes very well with other derivatives of the root verb. For instance, a snake is not something beautiful to look at and people do not await its appearance Nevertheless, when it is seen everybody runs and the news spreads very quickly. Thus a snake is also called *al-hilāl.*

Hamza Yusuf's selective insistence upon the term's secondary and metaphorical sense as the primary meaning is problematical. For example, although he quotes Ibn Manzur's *Lisān al-ʿArab,* "the *hilāl* got its name from the cry of joy that those who saw it uttered upon seeing it,"[48] he draws a conclusion contradicts that of this scholar. Ibn Manzur, representing the overwhelming majority, states that the *hilāl* acquired its name from the cry of joy made upon seeing the new moon, not from the act of "seeing" the new moon or its light. Therefore, Hamza's conclusion, "Clearly, according to the above evidence, the crescent moon is something that is seen. It is a physical phenomenon that upon seeing it people tend 'to cry for joy,' which is another accepted meaning of *ahalla/ yuhillu*"[49] is mistaken.

Moreover, his translation of al-Khalil ibn Ahmad's statement in *Al-ʿAyn* also reflects his selective understanding of the term, although it was not necessarily the original intent of this scholar, who was a very scientifically minded and precision-oriented philologist.

Al-Khalil writes:

هلّ السّحابُ بالمطر هلًّا، وآنهلّ بالمر انهلالًا، وهو شدّةُ آنصبابِه، ويَتَهَلَّلُ السّحابُ ببَرْقِهِ أي: يتلألأ. ويتهلّلُ الرّجلُ فرحاً. قال: تراه إذا ما جئتَهُ مُتَهلّلا ... كأنَّكَ تُعْطيهِ الّذي أنت سائلُهْ .[50]

(*Halla*) The rain poured down from the cloud – *hallan* – a real downpour… meaning the intensity of its gush; the cloud gleamed with its lightning: meaning it shone or flashed; and the man beamed with joy as [the poet said]: "You see him overjoyed when you come to him with a request as if you are at the giving end of what you actually are asking of him." (This means that the person alluded to is extremely generous and enjoys giving).

He continues:

والاستهلالُ: الصّوت. وكل مُتهلّلٍ رافع الصوت أو خافضه فهو مُهلّ ومُستَهِلّ. وأنشد: وأَلْفيتُ الخصومَ فهم لديه ... مُبْرْشِمَةٌ أهلّوا ينظرونا والهِلالُ: الحيّةُ الذّكر.[51]

And *al-istihlāl* is the voice. Every *mutahallil*, whether he shouts or lowers his voice, is called *muhill* and *mustahill*. It has been chanted: "And I found the enemy [defeated] nailed down by him, beginning to look at us [hysterically as if they were affixed or tacked to the earth]. *Al-hilāl* is also the male snake." (Here the meanings of voice, beginning and fear are emphasized.)

Bearing this multifaceted understanding in mind, let's now look at what he says about the new moon.

والهلالُ: غُرّةُ القَمَرِ حين يُهِلُّه النّاس في غُرّة الشّهر. يقال: أُهِلَّ الهلالُ ولا يُقال: هَل.[52]

Al-hilāl is the first day [beginning] of the moon [blaze, white spot, the finest part of the moon], when people rejoice over it [talk about it, see it] at the beginning of the month. It is said, *uhilla al-hilāl*, not *halla*.

Clearly, al-Khalil's statement does not confine the meanings of *al-hilāl* to mere naked-eye sighting of the new moon. A thorough study of his treatment of the term proves that he ascribed the following meanings to the verb *halla – yahillu*, among them rejoicing, announcing or raising the voice, beginning, shining, gleaming, and pouring down with intensity. Therefore, Hamza Yusuf's translation of his above-quoted statement as "when people actually see the crescent at the outset of the month" appears to be a selective insistence upon certain meanings at the expense of others.

He writes:

In determining the crescent moon, an important question arises: what exactly does "crescent" (*hilāl*) mean in the classical Arabic language? Furthermore, does our modern understanding of this word differ from the Arab understanding of the seventh century? The earliest and one of the most authoritative lexicons in the Islamic tradition is that of the linguist, al-Khalil b. Ahmad, of Oman. His book, *al-ʿAyn*, is the first scientific lexicon in human history. In it, he defines the word "crescent" (*hilāl*) as, "The first light of the moon, when people actually see the crescent at the outset of a month.... It is said, 'The crescent was seen' (*uhilla l-hilāl*) and not 'The crescent appeared' (*halla l-hilāl*).[53]

IMAM IBN TAYMIYYAH AND THE MEANINGS OF *HILĀL*

Ibn Taymiyyah clarifies this point:

وَذَلِكَ أَنَّ الْهِلالَ أَمْرٌ مَشْهُودٌ مَرْئِيٌّ بِالابْصَارِ. وَمِنْ أَصَحِّ الْمَعْلُومَاتِ مَا شُوهِدَ بِالابْصَارِ وَلِهَذَا سَمَّوْهُ هِلالا؛ لانَّ هَذِهِ الْمَادَّةَ تَدُلُّ عَلَى الظُّهُورِ وَالْبَيَانِ: إِمَّا سَمْعًا وَإِمَّا بَصَرًا كَمَا يُقَالُ: أَهَلَّ بِالْعُمْرَةِ وَأَهَلَّ بِالذَّبِيحَةِ لِغَيْرِ اللَّهِ إِذَا رَفَعَ صَوْتَهُ وَيُقَالُ لِوَقْعِ الْمَطَرِ الْهَلَلُ. وَيُقَالُ: اسْتَهَلَّ الْجَنِينُ إِذَا خَرَجَ صَارِخًا. وَيُقَالُ: تَهَلَّلَ وَجْهُهُ إِذَا اسْتَنَارَ وَأَضَاءَ. وَقِيلَ: إِنَّ أَصْلَهُ رَفْعُ الصَّوْتِ. ثُمَّ لَمَّا كَانُوا يَرْفَعُونَ أَصْوَاتَهُمْ عِنْدَ رُؤْيَتِهِ سَمَّوْهُ هِلالا.[54]

[It is called *hilāl*] because the new moon is a matter witnessed and sighted with the eyes. Since the information obtained from the eyes is the most accurate, that is why [the new sighted moon] is called a *hilāl*. [This is] because its root word leads to appearance [conspicuousness] and announcement (manifestation), either from listening or from sighting, for it is said: *ahalla* for umrah and *ahalla* for the slaughtered animal [raised one's voice to chant the name of other than Allah at the time of slaughter], meaning that [he] raised his voice. Pouring rain is called *al-halal*, and when the newborn baby comes out crying, it is referred to as *istahalla*. It is said: his face *tahallala* when the face gleams and shines. It is said that the origin of this [root] is raising the voice, and since the people used to raise their voices on sighting the new moon, they called [the new moon] a *hilāl*.

Al-Tabari also states that *ihlāl* is an "announcement":

والاهلال: رفع الصوت، يقال: أهل بكذا، أي رفع صوته.[55]

Ibn Hajar al-Asqalani asserts that the new moon was called *al-hilāl* because people used to raise their voices [announce their sighting] upon sighting the new moon.

قَالَ الطَّبَرِيُّ: الاهْلال هُنَا رَفْع الصَّوْت بِالتَّلْبِيَةِ وَكُلّ رَافِع صَوْته بِشَيْءٍ فَهُوَ مُهِلّ بِهِ، وَأَمَّا أَهَلَّ الْقَوْم الْهِلال فَأَرَى أَنَّهُ مِنْ هَذَا لانَّهُمْ كَانُوا يَرْفَعُونَ أَصْوَاتهمْ عِنْدَ رُؤْيَته.[56]

It should be clear by now that *hilāl's* original meanings are connected with raising the voice, loud rejoicing, the beginning of something, and so on, not with the shining light of the new moon. The new moon was called *hilāl* because it was the first sure sign of the new month and because those who saw it raised their voices to inform others about the new month's arrival. The interpretation of the new moon being sighted as opposed to being announced was based upon customary Arab use (*ʿurf*), which gradually took over this word's original linguistic meaning.[57] Since there was no accurate method other than naked-eye sighting, as Ibn Taymiyyah has mentioned in a quote above, the Arabs of the Propeht's time defined the new moon as something "seen" rather than "known." Moreover, as this phenomenon happened only during the first few days of the new month, the new moon was called *hilāl.* If this word had designated the new moon because of its light or shine, then the full moon would have had more right to be called *hilāl* because it is shinier, contains more light, and is seen by more people than the crescent. The moon toward the end of the month, specifically during the last two nights, has always been called *al-hilāl* because the Arabs used to talk about the end of the month and the arrival of a new month. Naked-eye sighting was just a means to ascertain the beginning and end of a lunar month.

Linguistically, the astronomically calculated new moon (even without light) could, in principle, be called *hilāl* if its beginning (arrival) could be clearly ascertained via calculations and announced so that people could raise their voices about its arrival and talk about it. Such a definition would cover almost all of the original aspects and attendant corollaries of the root word *halla*. Undoubtedly, the pre-modern Arabs used naked-eye sighting as the most reliable technique and mode of ascertaining the new month's beginning and end of a new month, as Ibn Taymiyyah stated, and that is why they called it *hilāl.*

I am not asserting here that Ibn Taymiyyah accepted the use of calculations regarding this matter. In fact he rejected this approach on the grounds that calculations were not precise enough to be employed in such an important devotional matter and that they were the products of astrology and fortune telling, both of which were prohibited professions that conflicted directly with Islam's pristine *ʿaqīdah* (belief system). All I am saying here that the cultural meanings of *al-hilāl* gradually became so customary that people started equating them with the original linguistic meanings. Now, if we can attain the same degree of certainty from calculations and announce the arrival of the new month and new moon, we can possibly call it *hilāl.*

Meanings of *al-ʿUrf*

Some jurists maintain that Allah has commanded us to accept the *ʿurf* (customs of people) as long as they do not contradict an established Islamic text, such as consuming intoxicants. They substantiate their position with the following verse:

(خُذِ الْعَفْوَ وَأْمُرْ بِالْعُرْفِ وَأَعْرِضْ عَنِ الْجَاهِلِينَ.) (الأعراف: 199)

[O Prophet]: "Hold on to forgiveness, command what is known [good, customary] and ignore the ignorant." (7:199)

(خُذْ الْعَفْوَ وَأْمُرْ بِالْعُرْفِ): إنَّ مَعْنَى الْعُرْفِ: كُلُّ مَا عَرَفَتْهُ النُّفوسُ مِمَّا لا تَرُدُّهُ الشَّرِيعَةُ، قَالَ ابْنُ ظَفرٍ فِي الْيَنْبُوعِ "الْعُرْف" مَا عَرِفَهُ الْعُقَلاءُ بِأَنَّهُ حَسَنٌ.[58]

Adopt forgiveness and command *ᶜurf*. *ᶜUrf* means "whatever is approved of by the people and is not prohibited by the Shariᶜah." Ibn Zafar, in *Al-Yanbu'a*, states: "*ᶜUrf* is what the thoughtful people consider to be good."

The exegete Ibn Ghalib ibn ᶜAtiyyah[59] explains the verse as: "[C]ertainly *al-ᶜurf* here means those practices to which people became accustomed from among those practices which are not rejected by the Shariᶜah."

The Prophet accepted and incorporated numerous *jāhiliy* customs, such as the *diyah* (indemnity for bodily injury or blood money) of one hundred camels, *qasāmah* (making the male members of a clan responsible for blood money), ligiting a fire at Muzdalifah to guide the pilgrims, and proceeding from Arafah. Even some verses and rulings gave great weight to local customs when promulgating social rules. For instance, the Qur'an requires slaves and pre-pubescent children to seek permission before entering a house at three specific times: before *fajr*, during the afternoon siesta when "you take off your clothes" and after *ᶜishā'*.

(يا أيها الذين آمنوا ليستأذنكم الذين ملكت أيمانكم والذين لم يبلغوا الحلم منكم ثلاث مرات من قبل صلاة الفجر وحين تضعون ثيابكم من الظهيرة ومن بعد صلاة العشاء ثلاث عورات لكم ليس عليكم ولا عليهم جناح بعدهن طوافون عليكم بعضكم على بعض كذلك يبين الله لكم الآيات والله عليم حكيم.) (النور: 58)

O you who believe! Let your slaves, and those of you who have not reached puberty, ask leave of you on three occasions [before they come into your presence]: Before the dawn prayer, and when you take off your clothes during the heat of the afternoon, and after the night prayer. These are the three periods of privacy for you. It is no sin for them or for you at other times, when some of you go round to attend upon others [if they come into your presence without leave]. Thus Allah makes clear the revelations for you. Allah is the Knower, Wise. (24:58)

This social rule was based upon a custom of Madinah's Arab/Muslim community. The ruling of "taking off one's clothes off time during the afternoon" would change along with any change in the people's customary practice.

One established *fiqhī* rule requires a jurist to try to look at the reasoning behind a ruling and not just accept its literal sense. The rule could have been based upon a custom *ʿurf.*

الْوَصْفُ الْمُعَلَّلُ بِهِ قَدْ يَكُونُ عُرْفِيًّا.[60]

The reasoning factor [behind a rule] may be a custom (*ʿurf*).

Another *fiqhī* rule explains that if a ruling on a given subject is based upon *ʿurf,* then that ruling will change along with any change in that particular custom. Of course, any rulings founded upon such unequivocal Islamic texts as the Qur'an and the Sunnah are immutable, even if the customs are in conflict with them.

إنَّ الاحْكَامَ الَّتِي تَتَغَيَّرُ بِتَغَيُّرِ الازْمَانِ هِيَ الاحْكَامُ الْمُسْتَنِدَةُ عَلَى الْعُرْفِ وَالْعَادَةِ؛ لانَّهُ بِتَغَيُّرِ الازْمَانِ تَتَغَيَّرُ احْتِيَاجَاتُ النَّاسِ، وَبِنَاءً عَلَى هَذَا التَّغَيُّرِ يَتَبَدَّلُ أَيْضًا الْعُرْفُ وَالْعَادَةُ وَبِتَغَيُّرِ الْعُرْفِ وَالْعَادَةِ تَتَغَيَّرُ الاحْكَامُ ...، بِخِلافِ الاحْكَامِ الْمُسْتَنِدَةِ عَلَى الادِلَّةِ الشَّرْعِيَّةِ الَّتِي لَمْ تُبْنَ عَلَى الْعُرْفِ وَالْعَادَةِ فَإِنَّهَا لا تَتَغَيَّرُ.[61]

The rulings/injunctions that change as times change are those which are based on custom (*ʿurf*). This is because, as times change, so do people's needs, and, based on these changes, the custom (*ʿurf*) changes too. As a result, the rulings change. Nevertheless, text-based Islamic rulings that are not originally based upon custom (*ʿurf*) do not change.

It is pertinent to note here that Abu Yusuf, the celebrated student of Abu Hanifah and the fundamental authority on Hanafi fiqh, maintains that if there is a conflict between the established text and the people's customs, then it is essential to determine whether or not the original text was based upon previous customs and norms. In other words, existing customs are to be preferred over an established text (*naṣṣ*) if the text was based upon an earlier custom or norm.

عَلَى أَنَّ الامَامَ أَبَا يُوسُفَ يَقُولُ: إذَا تَعَارَضَ النَّصُّ وَالْعُرْفُ يُنْظَرُ فِيمَا إذَا كَانَ النَّصُّ مَبْنِيًّا عَلَى الْعُرْفِ وَالْعَادَةِ أَمْ لا؟ فَإِذَا كَانَ النَّصُّ مَبْنِيًّا عَلَى الْعُرْفِ وَالْعَادَةِ تُرَجَّحُ الْعَادَةُ وَيُتْرَكُ النَّصُّ.[62]

Imam Abu Yusuf states: If there is a conflict between the text and the custom (*ʿurf*), then it is necessary to check whether the text is based on custom (*ʿurf*) or not. If the text is based upon custom (*ᶜurf*), then the [current] custom (*ᶜurf*) is preferred and the text is abandoned.

I conclude this part of the discussion by noting that Allah never used a word that required naked-eye sighting of the new moon of Ramadan or Dhu al-Hijjah in order to fast Ramadan or perform hajj. Had doing so been so important, He could have used such categorical terms as "whoever sights the new moon of Ramadan and Dhu al-Hijjah" and then granted exemption to those who were sick or traveling, for instance. But instead, He used a phrase like "whoever witnesses the month," which means whoever knows the month of Ramadan and is healthy and resident, in order to leave the door open for the event to be seen or the news to be heard, not to mention other possible future interpretations.

The Qur'an, the eternal and literal Word of God, responds to peoples of all times. The phrase analyzed above permitted pre-modern Muslims to use the most reliable method of witnessing the month of Ramadan (viz., naked-eye sighting), and is flexible enough to resolve our current uncertainties and confusion. Should physical sighting be impossible, calculations based upon the "meticulously calculated" stages of the moon can be used. After all, the ultimate goal is certainty about the month, not its physical sighting.

The Prophet commanded Muslims to start and end Ramadan according to naked-eye sighting because of the existent constraints of his time. In the final analysis, it is clear that this was not the only method he used to attain certainty in this matter. For instance, he gave us the alternative of completing thirty days of Ramadan or Shaᶜban in the event of such natural obscurities as clouds. He never required us to look for the new moon on 30 Shaᶜban or 30 Ramadan because the new month would certainly start after the thirtieth day. Such an approach proves that certainty, not actual sighting, is the prerequisite for fasting Ramadan.

Any detailed discussion of the relevant prophetic narrations will prove that the Prophet required naked-eye sighting only if we were to start Ramadan after 29 Shaᶜban or start Shawwal after fasting only twenty-nine days during Ramadan. Since naked-eye sighting was the only method that gave certainty, he required its use on the twenty-ninth day of the month. His accompanying words, "We are an unlettered people; we neither write nor calculate," pinpointed the real reason for this requirement.

By his time, the Jews had fully adopted the use of calculations or intercalations to synchronize their lunar calendar with the solar calendar. The pre-Islamic Arabs followed this innovation of adding days to lunar months and an extra thirteenth month to the twelve-month lunar year. As this caused hajj and other sacred rituals to occur not at their divinely prescribed times, but at the times suited to the people's financial and political needs, the Prophet restored the original schedule and reconnected the months to the new moon's arrival.

The Qur'an addresses intercalation in the following verses.

(إن عدة الشهور عند الله اثنا عشر شهرا في كتاب الله يوم خلق السماوات والأرض منها أربعة حرم ذلك الدين القيم فلا تظلموا فيهن أنفسكم وقاتلوا المشركين كافة كما يقاتلونكم كافة واعلموا أن الله مع المتقين إنما النسيء زيادة في الكفر يضل به الذين كفروا يحلونه عاما ويحرمونه عاما ليواطئوا عدة ما حرم الله فيحلوا ما حرم الله زين لهم سوء أعمالهم والله لا يهدي القوم الكافرين.) (التوبة: 36-37)

Behold, the number of months, in the sight of God, is twelve months, [laid down] in God's decree on the day when He created the heavens and the earth; of these, four are sacred: this is the ever true law of God. Do not, then, sin against yourselves with regard to these [months]... The intercalation [of months] is but one more instance of [their] refusal to acknowledge the truth: [a means] by which those who are bent on denying the truth are led astray. They declare this [intercalation] to be permissible in one year and forbidden in [another] year, in order to conform [outwardly] to the number of months that God has made sacred: and thus they make lawful what God has forbidden. Goodly seems unto them the evil of their own doings, since God does not grace with His guidance people who refuse to acknowledge the truth. (9:36–37)

According to most exegetes, these verses refer to the pre-Islamic Arabs' arbitrary intercalations as regards the lunar months and the lunar years.[63] Al-Razi elaborates upon this point:

إن القوم علموا أنهم لو رتبوا حسابهم على السنة القمرية، فإنه يقع حجهم تارة في الصيف وتارة في الشتاء، وكان يشق عليهم الأسفار ولم ينتفعوا بها في المرابحات والتجارات، لأن سائر الناس من سائر البلاد ما كانوا يحضرون إلا في الأوقات اللائقة الموافقة، فعلموا أن بناء الأمر على رعاية السنة القمرية يخل بمصالح الدنيا، فتركوا ذلك واعتبروا السنة الشمسية، ولما كانت السنة الشمسية زائدة على السنة القمرية بمقدار معين، احتاجوا إلى الكبيسة وحصل لهم بسبب تلك الكبيسة أمران: أحدهما: أنهم كانوا يجعلون بعض السنين ثلاثة عشر شهراً بسبب اجتماع تلك الزيادات. والثاني: أنه كان ينتقل الحج من بعض الشهور القمرية إلى غيره، فكان الحج يقع في بعض السنين في ذي الحجة وبعده في المحرم وبعده في صفر.[64]

> The [pre-Islamic Arab] people knew that following the lunar year would cause the hajj to fall sometimes during the summer and sometimes during the winter. It was hard for them to travel and this [movement of the hajj between seasons] was not financially profitable...Consequently, they replaced the lunar year with a solar year. They had to use intercalation because the [days in a] solar year were more than [the days in a] lunar year. This intercalation resulted in two problems. Firstly, they had to add a 13th month owing to (the synchronization) with the extra days of the solar year. Secondly, the hajj had to revolve between various lunar months. The hajj would fall in some years in the actual month of Dhu al-Hijjah and in other years in the month of al-Muharram or Safar.

Given this reality, Allah warned them and declared their moving of hajj from the sacred months to other ordinary months as tantamount to *kufr* (disbelief), as al-Razi explains:

فلهذا السبب عاب الله عليهم وجعله سبباً لزيادة كفرهم، وإنما كان ذلك سبباً لزيادة الكفر، لأن الله تعالى أمرهم بإيقاع الحج في الأشهر الحرم، ثم إنهم بسبب هذه الكبيسة أوقعوه في غير هذه الأشهر.[65]

The pre-Islamic Arabs, who used to postpone, name, and rename the months in accordance with their political, economic, and military situations, had no scientific way to ascertain the exact time of when for the moon's, as well as the sun's, rising and setting. In addition, there is no proof that Madinah's Jewish community had such developed and precise mathematical calculations. As will be discussed later on, the Jews' calculations were based upon a fixed average lunar month rather than any astronomical calculation of actual moon births. Tracy R. Rich clarifies this point:

> Note that the calculated *molad* (birth) does not necessarily correspond precisely to the astronomical new moon. The length of time from one astronomical new moon to the next varies somewhat because of the eccentric orbits of the Earth and moon; however, the moladot of Rabbi Hillel's calendar are set using a fixed average length of time: 29 days, 12 hours, and 793 parts (or in Hebrew, *chalakim*). The amount of time is commonly written in an abbreviated form: 29d 12h 793p.[66]

This was the main reason why the Qur'an and the Prophet rejected the Jewish approach to the lunar months. The Qur'an discarded such intercalations and restored the calendar to its original form, a point clearly emphasized by the Prophet in his last sermon: "Today, certainly the calendar has returned to its original form as God had created it to be at the times of creation of the heavens and the earth." Al-Razi states that by doing so, the Prophet restored the sacred months to their original schedule.

(ألا إن الزمان قد استدار كهيئته يوم خلق السموات والأرض السنة إثنا عشر شهراً) وأراد أن الأشهر الحرم رجعت إلى مواضعها.[67]

"Certainly the time has taken its original form as it was meant to be at the time of creation of the heavens and ẹarth, the year is twelve months." He meant that the sacred months have returned to their (original) right timings.

All of the prophetic hadiths that stress naked-eye sighting for confirming or negating the month of Ramadan must be understood against this background. The prophetic insistence upon this method sought to restore the sacred months to their original schedule. In fact, his statement that "We are an unlettered people; we neither write nor calculate" clearly refers to the arbitrary intercalations discussed above, not to the calculations produced by modern-day experts. The Qur'anic argument actually supports the latter type of calculations by stating that Allah has fixed the moon's phases so that people will have an accurate basis upon which to derive a precise calendar.

THE *MAWĀQĪT* ARGUMENT

The following Qur'anic verse is frequently quoted to prove that the Qur'an demands the actual sighting of the moon to confirm the lunar months. This verse in reality is refuting the arbitrary intercalations and has little connection with the subject of moon sighting or astronomical calculations. The Qur'an states:

(يَسْأَلُونَكَ عَنِ الأَهِلَّةِ قُلْ هِيَ مَوَاقِيتُ لِلنَّاسِ وَالْحَجِّ وَلَيْسَ الْبِرُّ بِأَنْ تَأْتُوا الْبُيُوتَ مِنْ ظُهُورِهَا وَلَكِنَّ الْبِرَّ مَنِ اتَّقَى وَأْتُوا الْبُيُوتَ مِنْ أَبْوَابِهَا وَاتَّقُوا اللَّهَ لَعَلَّكُمْ تُفْلِحُونَ.) (البقرة: 189)

They ask you about the (*ahillah*) new moons; say: "They are appointed times for the people and for the hajj. It is not righteousness to enter houses through the back doors; righteousness consists of *taqwā*. So enter houses through their [front] doors and fear Allah so that you may attain felicity." (2:189)

In the above verse, the Qur'an scolds the pre-Islamic Arabs for their misplaced priorities. For example, they used to enter or leave their houses or tents through a hole in the back or down a ladder after putting on their hajj garments (*iḥrām*). Allah rebukes them for worrying so much about that which is external to the hajj at the expense of its true timings and essence.

Hamza Yusuf recognizes this context by quoting al-Qurtubi:

> Hajj is specifically mentioned in this verse [as opposed to Ramadan] because it is among the devotional months; the knowledge of its commencement is essential, and [determining it using] intercalation is not permissible, as doing so displaces it from its proper time. *This is in opposition to what the Arabs considered acceptable at the time, as their practice was to perform hajj based upon calculation (ʿadad) and alteration of the months. Thus, God nullified their words and deeds.*[68]

He also realizes that this divine command came "in preparation for the final prohibition on determining the hajj based upon intercalation and not sighting, as the pre-Islamic Arabs sometimes predetermined hajj, so they did not need to look for the moon during the hajj season."[69]

Therefore, we conclude that this verse prohibits only the arbitrarily forced intercalations. We also reiterate that naked-eye sighting was the only known method that gave certainty that the new month had arrived. This is why the Prophet told Muslims to use it or complete thirty days, instead of using intercalation to synchronize the lunar year with the solar year in order to derive financial and other worldly gains. He wanted to ascertain that Ramadan was observed during its sacred time. Given that naked-eye sighting was the only available method to attain certainty and not a goal in itself, any type of calculations (such as those of our own time) that provides the same level of certainty and do not compromise the true timing of Ramadan and Dhu al-Hijjah, are permissible.

MOON-SIGHTING SUPPLICATIONS ARE BASED UPON WEAK REPORTS

It is argued that some hadith compilations contain multiple supplications that are to be recited upon sighting the new moon. Their existence and the prophetic recommendation that they be recited at that particular time are used as arguments for suggesting (or even demanding) the necessity of naked-eye sighting.

While it is true that several hadith sources do narrate different formulas for such supplications, al-Bukhari and Muslim did not consider these reports worthy of their attention, mainly due to the serious doubts about their authenticity. Abu Dawud narrated two conflicting reports: one describes how the Prophet, at the first glance, would turn his face away from the new moon; the other states that he used to recite some form of supplication. In the conclusion of his narrations on the subject, Abu Dawud affirms that there is no relevant authentic report from the Prophet. His two conflicting reports are given below:

حَدَّثَنَا مُوسَى بْنُ إِسْمَعِيلَ حَدَّثَنَا أَبَانُ حَدَّثَنَا قَتَادَةُ أَنَّهُ بَلَغَهُ أَنَّ النَّبِيَّ صَلَّى اللَّهُ عَلَيْهِ وَسَلَّمَ كَانَ إِذَا رَأَى الْهِلَالَ قَالَ هِلَالُ خَيْرٍ وَرُشْدٍ هِلَالُ خَيْرٍ وَرُشْدٍ هِلَالُ خَيْرٍ وَرُشْدٍ آمَنْتُ بِالَّذِي خَلَقَكَ ثَلَاثَ مَرَّاتٍ ثُمَّ يَقُولُ الْحَمْدُ لِلَّهِ الَّذِي ذَهَبَ بِشَهْرِ كَذَا وَجَاءَ بِشَهْرِ كَذَا.[70]

Musa ibn Ismaʿil reported on the authority of Aban on the authority of Qatadah that the Prophet at seeing the new moon would say, "a new moon of goodness and guidance (thrice), I believe in the One who created you (thrice), then he will say, "thanks be to Allah who took such and such month away and brought forth such and such month."

Ahmad ibn Hanbal narrates that:

حدثنا عبد الله حدَّثني أبي ثنا أبو عامر ثنا سليمان بن سفيان المدايني حدَّثني بلال بن يحيى بن طلحة بن عبيد الله عن أبيه عن جده: (أن النبي صلى الله عليه وسلم كان إذا رأى الهلال قال: اللهم أَهِلَّه علينا باليمن والإيمان والسلامة والإسلام ربي وربك الله).[71]

Upon seeing the crescent moon, the Prophet would say: "O God, cause this new moon to come upon us in safety and sound faith, security and submission. My Lord and your Lord is Allah."

Al-Tirmidhi, Ibn Habban, and others narrate this hadith. Abu Dawud narrates that the Prophet used to turn his face away upon sighting the new moon.

حدثنا مُحَمَّدُ بنُ الْعَلَاءِ أَنَّ زَيْدَ بنَ حُبَابٍ أخبرهم عن أَبِي هِلَالٍ عن قَتَادَةَ: (أَنَّ رَسُولَ الله صلى الله عليه وسلم كَانَ إِذَا رَأَى الْهِلَالَ صَرَفَ وَجْهَهُ عَنْهُ).[72]

Qatadah reports that the Prophet would turn his face away on sighting the new moon.

Abu Dawud concludes his chapter on "What a Person Should Say on Seeing the New Moon" by observing that none of the hadiths narrated in the chapter are authentic.

قَالَ أَبو دَاوُدَ: لَيْسَ عن النَّبِيِّ صلى الله عليه وسلم في هذَا الْبَابِ حَدِيثٌ مُسْنَدٌ صَحِيحٌ.[73]

Ibn al-ʿArabi, commenting on this question of supplicating upon seeing the new moon, shows the contradictory nature of these reports:

أَمَّا إِنَّهُ رَوَى أَبو دَاوُد وَغَيْرُهُ عَنْ قَتَادَةَ بَلاغًا عَنْ النَّبِيِّ -صَلَّى اللَّهُ عَلَيْهِ وَسَلَّمَ- حَدِيثَيْنِ مُتَعَارِضَيْنِ: أَحَدُهُمَا: (أَنَّ النَّبِيَّ - صَلَّى اللَّهُ عَلَيْهِ وَسَلَّمَ - كَانَ إِذَا رَأَى الْهِلالَ أَعْرَضَ عَنْهُ). الثَّانِي: (أَنَّهُ كَانَ إِذَا رَآهُ قَالَ: هِلالُ خَيْرٍ وَرُشْدٍ، آمَنْت بِالَّذِي خَلَقَك ثَلاثَ مَرَّاتٍ ثُمَّ يَقُولُ: الْحَمْدُ لِلَّهِ الَّذِي ذَهَبَ بِشَهْرِ كَذَا وَجَاءَ بِشَهْرِ كَذَا). قَالَ الْقَاضِي: وَلَقَدْ لُكْته فَمَا وَجَدْت لَهُ طَعْمًا. وَقَدْ أَخْبَرَنَا الْمُبَارَكُ بْنُ عَبْدِ الْجَبَّارِ، أَخْبَرَنَا ابْنُ زَوْجِ الْحُرَّةِ أَنْبَأَنَا النَّجِيُّ، أَنْبَأَنَا ابْنُ مَحْبُوبٍ، أَنْبَأَنَا ابْنُ سَوْرَةَ، أَنْبَأَنَا مُحَمَّدُ بْنُ بَشَّارٍ، أَنْبَأَنَا أَبو عَامِرٍ الْعَقَدِيُّ، أَنْبَأَنَا سُلَيْمَانُ بْنُ سُفْيَانَ الْمَدَنِيُّ، أَنْبَأَنَا بِلالُ بْنُ يَحْيَى بْنِ طَلْحَةَ بْنِ عُبَيْدِ اللَّهِ عَنْ أَبِيهِ عَنْ جَدِّهِ طَلْحَةَ بْنِ عُبَيْدِ اللَّهِ (أَنَّ النَّبِيَّ - صَلَّى اللَّهُ عَلَيْهِ وَسَلَّمَ - كَانَ إِذَا رَأَى الْهِلالَ قَالَ: اللَّهُمَّ أَهِلَّهُ عَلَيْنَا بِالْيُمْنِ (أَنَّ النَّبِيَّ - صَلَّى اللَّهُ عَلَيْهِ وَسَلَّمَ - كَانَ إِذَا رَأَى الْهِلالَ قَالَ: اللَّهُمَّ أَهِلَّهُ عَلَيْنَا بِالْيُمْنِ وَالإِيمَانِ وَالسَّلامَةِ وَالإِسْلامِ). قَالَ ابْنُ سَوْرَةَ: حَسَنٌ غَرِيبٌ.[74]

Abu Dawud and others have reported from the Prophet two contradictory reports. Firstly, the Prophet used to turn his face away at the first glance of a new moon. Secondly, that he would say, "a new moon of goodness and guidance [thrice], I belive in the One who created you [thrice]," then he will say, "thanks be to Allah who took such and such month away and brought forth such and sunc month." Al-Qadi said, "I did not find this [second] report authentic at investigation." ... Ibn Surah also found the following report as "*gharīb*" [its chain of narrators is not authentic]. "O Allah bless us with good luck, faith, protection and Islam during this month."

Even al-Tirmidhi, after narrating the hadith on these supplications, states that this is a "good but strange" hadith.

رواه الترمذي وقال هذا حديث حسن غريب.[75]

It is a well-known principle in hadith science that a weak hadith is accepted in matters of virtues (*faḍā'il*) but not in stipulating rules. This is why these supplications, which were originally based upon weak hadiths, have a wide circulation among Muslims. Although I believe that reciting these supplications upon sighting the new moon is rewarding, such weak hadiths cannot prove that naked-eye sighting is a prerequisite for fasting Ramadan. This reward can be achieved whenever one sees the new moon. In any case, not everyone can see the new moon on the first day. Such supplications can be recited whenever, during the first seven days of the new month, one sees the new moon, as the metaphorically oriented, linguistic meanings of *al-hilāl* denote. Reciting such supplications are irrelevant to claiming that the month of Ramadan is solely dependent upon naked-eye sighting. Moreover, it is recommended that such supplications be made during all months, for the Prophet reportedly said that Muslims should recite them whenever they see the new moon. Should we stop doing so just because many people no longer engage in naked-eye sighting? The reward for doing so is granted and the Sunnah is fulfilled whenever someone recites them during the appropriate time. In no way does it prove that naked-eye sighting is the precondition for fasting Ramadan.

CHAPTER 3

THE SUNNAH IN REALITY REQUIRES "CERTAINTY": THE LEGAL CAUSE IS THE "COMING OF RAMADAN"

Traditionally it has been argued that naked-eye sighting is the cause (*sabab*) and that Ramadan can be determined only by fulfilling it, and that the prophetic reports established only two methods to suitable methods of attaining certainty in this matter: naked-eye sighting or completing thirty days (*ikmāl*). However, one question remains: is naked-eye sighting a precondition for starting Ramadan immediately after 29 Shaʿban only or it is also a requirement on 30 Shaʿban?

It is generally agreed that naked-eye sighting is not required in the latter case. In fact, many Muslim countries announce that Ramadan will start on the thirty-first day if the new moon has not been sighted on the twenty-ninth day of the month. Looking for the actual moon on the thirtieth day is only a recommended *sunnah*. This entire debate regarding naked-eye sighting as a prerequisite to fasting or not fasting is related only to the twenty-ninth day of the month. Consequently, such a sighting could be a legal reason for starting to fast after 29 Shaʿban, not 30 Shaʿban, for the cause and effect go hand in hand. The prophetic hadith requiring the completion of thirty days comes into effect *only if the skies are cloudy*.

In other words, if naked-eye were the sole cause of starting or ending Ramadan, then this cause would have been required on both 29 Shaʿban and 30 Shaʿban. As the latter is not the case we can conclude that certainty, not naked-eye sighting, of Ramadan's arrival is a precondition for the fast. In addition, it can be surmised that the prophetic traditions dealing with naked-eye sighting are concerned primarily with starting or ending the month on 29 Shaʿban. Had they been connected with fasting after 30 Shaʿban, then naked-eye sighting would have been required even on the evening of that particular day.

Mokhtar Maghraoui argues that "do not fast until you see the new moon" is categorical in demanding that sighting and only sighting is the sole means to ascertain the beginning of the month.[1] But why would the Prophet insist upon naked-eye sighting as the condition for fasting only on the evening of 29 Shaʿban but not on the following evening? If it were intrinsically obligatory to do so, it would have been required on both days. A binding legal cause cannot be established as a prerequisite objective for only one day of the month and not the other day. The cause and effect go hand in hand in all situations, as Abu Ishaq al-Shatibi has stated.

ما أثبت سبباً، فهو سبب أبداً لا يرتفع...[2]

The established [legal] cause always remains the cause and never changes.

On the other hand, if we establish certainty about the new month's arrival as the real reason for the prophetic insistence upon naked-eye sighting on 29 Shaʿban, we can fully understand why he did not require it on the evening of the following day. It is also very important to understand this prophetic insistence within the context of the challenges and confusion caused by arbitrary intercalations and illiteracy.

Rashid Rida has observed this fact in quote given below: the Lawgiver requires naked-eye sighting as a means to determine the sacred timings fixed for acts of worship (e.g., Ramadan), not to worship the act of sighting in itself. The Prophet referred to naked-eye sighting and completion of thirty days because the Ummah was unlettered. The objective of his prophethood was to bring the Ummah out of its unlettered state, not to push it further into illiteracy.

(العلم بهذه الأوقات، وليس التعبّد برؤية الهلال... وما ذكره صلى الله عليه وسلم من نوط إثبات الشهر برؤية الهلال أو اكمال العدّة بشرطه، قد علّله بكون الأمّة في عهده كانت أميّة، ومن مقاصد بعثته إخراجها من الأميّة لا إبقاؤها فيها...).[3]

The Lawgiver's intent is the knowledge of these timings and not worship through moon sighting ... The Prophet had connected confirmation by actual moonsighting or completion with the condition that the Ummah was unlettered during his times. The goal of his prophethood has been to bring the Ummah out of its unlettered status and not to continue stustaining it.

HADITHS ON SIGHTING THE MOON

I now turn to the original texts of the prophetic injunctions (hadiths) already mentioned, many of which clearly command Muslims to start and end Ramadan via naked-eye sighting. They must be understood in the light of our recent discussion, namely, that the Prophet emphasized this method in an effort to restore the lunar calendar to its original form. The pre-Islamic Arabs fixed these months with the help of artificial intercalations to harmonize them with their solar calendar solely for their own benefit. The Prophet commands:

حدّثنا آدمُ حدَّثَنا شُعبةُ حدَّثنا محمدُ بنُ زيادٍ قال: سمعتُ أبا هُريرةَ رضيَ اللّهُ عنهُ يقول: قال النبيُّ صلى الله عليه وسلم ــ أو قال: قال أبو القاسم صلى الله عليه وسلم ــ (صُوموا لِرُؤْيتهِ وأفطِروا لرُؤيته، فإن غُبِّيّ عليكم فأكملوا عِدَّةَ شَعبانَ ثلاثين).[4]

The Prophet said, "Fast on sighting it [the new moon] and break the fast on sighting it. Complete thirty days of Shaʿban if it is cloudy."

وحدّثنا عُبَيْدُ اللّهِ بْنُ مُعاذٍ. حَدَّثَنا أَبِي. حَدَّثَنا شُعْبَةُ عَنْ مُحَمَّدِ بْنِ زِيَادٍ قَالَ: سَمِعْتُ أَبا هُرَيْرَةَ، رَضِيَ اللّهُ عَنْهُ يَقُولُ: قَالَ رَسُولُ اللّهِ (صُومُوا لِرُؤْيَتِهِ وَأَفْطِرُوا لِرُؤْيَتِهِ. فَإِنْ غُمِّيَ عَلَيْكُمُ الشَّهْرُ فَعُدُّوا ثَلاَثِينَ).[5]

The Prophet said, "Fast on sighting it [the new moon] and break the fast on sighting it. Count thirty days if the month is concealed from you [the sky being cloudy]."

Other narrations use negative formulas to emphasize the same point:

حدّثنا عبدُ اللّهِ بنُ مَسْلمةَ عن مالكٍ عن نافعٍ عن عبدِ اللّهِ بنِ عُمَر رضيَ اللّهُ عنهما: (أنَّ رسولَ اللّهِ صلى الله عليه وسلم ذَكَرَ رَمضان فقال: لا تَصوموا حتّى تَرَوُا الهلالَ، ولا تُفْطِرُوا حتّى تَرَوْه، فإن غُمَّ عليكم فاقدُروا له).[6]

The Prophet mentioned Ramadan and said, "Do not fast until you see the moon and do not break the fast until you see it. Estimate it when it is cloudy."

Imam Ahmad reports the same:

حدثنا عبد الله، حدثني أبي، حدثنا عبد الرحمن، حدثنا مالك، عن نافع، عن ابن عمر، عن النبي صلى الله عليه وسلم قال: (لا تصوموا حتى تَروا الهلال، ولا تفطروا حتى تَرَوْه، فانْ غُمَّ عليكم فاقْدُرُوا له).[7]

Do not fast until you see the moon and do not break the fast until you see it. Estimate it when it is cloudy.

The Problem

Hamza Yusuf, discussing the nature of the problem, observes:

> The Islamic calendar is lunar. Lunar calendars follow the phases of the moon, beginning with the crescent moon and ending with the conjunction of the moon and the sun in their respective perceived movements around the earth. The time of one lunation or complete cycle of the moon in a lunar month is approximately 29.5 days. This must be averaged because the moon does not travel at a constant speed nor does it travel in a perfect circle but in an elliptical orbit around the earth. The moon's monthly cycle around the earth varies between 29.2 days and 29.8 days, which means that throughout the year there will be approximately six months in which there are twenty-nine days, and six months in which there are thirty days. The total number of days in a lunar year is approximately 354, which is eleven days shorter than the average solar year. *This results in the lunar year progressing through the fixed solar year rather than beginning and ending at the same time as the solar. In order to make the lunar years consistent with the solar, many pre-Islamic societies intercalated or added days to the lunar months. This enabled them to follow a lunar calendar without having it depart from the fixed seasons of the solar calendar. This was and remains the practice of the Jewish community, which intercalates a thirteenth month every three years in order to align the lunar calendar with the solar. The pre-Islamic Arabs used a lunar calendar but both calculated and intercalated their calendars when suitable for their needs. Their general practice however was to rely on a physical sighting of the crescent.*[8]

He continues,

> The Islamic lunar calendar is not to be tampered with, as the Prophet … prohibited intercalation in his farewell address to his community during the final pilgrimage. Islam condemns intercalation, regarding it as a rejection of the natural order inherent in the perfection of the lunar calendar that God has provided humanity for measuring their time. *For this reason, in a number of hadith that achieve the status of infallible* (mutāwatir) *and thus on par with the legislative authority of any verse in the Qur'an, the Prophet [s] commanded Muslims to base their month on the physical sighting of the new moon and stipulated that if not seen on the twenty-ninth completed day of the previous month on a clear evening, or if clouds or other atmospheric barriers hindered visibility, then to complete thirty days of the previous month and begin the new month the following sunset, which would mean on the thirty-first day following the previous sighting or a calculation of thirty days.*[9]

Accordingly, it is important to note the following:

First, the prophetic hadith clearly asks for naked-eye sighting as a means of certainty, not as a precondition for fasting. This type of sighting was required by the Prophet only because no other available method could confirm the new moon's presence, the sign of the new month's commencement. As he said: "We are an unlettered people. We neither write nor calculate."

The month sometimes consists of twenty-nine or thirty days. Acts of worship are connected with time, which in Islamic understanding is connected with the moon (not the sun), for the Islamic calendar is lunar. The Sharʿiah does not want us to start fasting before Ramadan has begun or to miss a day of it by celebrating the ʿId on the last day of Ramadan. That is why the Prophet told us not to start or finish this month a day or two ahead of time, bur rather to wait until certainty about the new moon's arrival has been obtained.

حدثنا قُتَيْبَةُ، حدثنا أبو الأحْوَصِ عن سِمَاكِ بنِ حَرْبٍ عن عِكْرِمَةَ عن ابنِ عباسٍ، قال: قال رسولُ الله (لا تَصُومُوا قَبْلَ رَمَضَانَ، صُومُوا لِرُؤْيَتِهِ وأفْطِرُوا لِرُؤْيَتِهِ، فإنْ حَالَتْ دُونَهُ غَيَابَةٌ فأكْمِلُوا ثلاثين يَوْماً).[10]

Qutaybah narrated from Abu al-Ahwas, from Simak ibn Harb, who reported that ʿAbd Allah ibn ʿAbbas said: "The Prophet said, 'Do not fast immediately before Ramadan. Start the fast with the sighting of the new moon and break your fast with the sighting of it. If the sky is overcast, then complete thirty days.'"

حدّثنا عبدُاللَّهِ بنُ سعيدٍ ثَنا إسماعيلُ بنُ عليةَ، ثَنا حاتمُ بنُ أبي صغيرةَ، عن سماكِ بنِ حربٍ، قالَ: أصبحتُ في يومٍ قد أشكلَ عليَّ من شعبانَ أو من شهرِ رمضانَ، فأصبحتُ صائماً فأتيتُ عكرمةَ فإذا هوَ يأكُلُ خبزاً وبقلاً، فقالَ: هلمَّ إلى الغداءِ فقلتُ: إني صائمٌ فقالَ: أُقْسِمُ باللَّهِ لَتُفْطِرَنَّ، فلما رأيتُهُ حلفَ ولا يستثني تقدَّمْتُ فعذّرت وإنما تسحرتُ قبيلَ ذَلِكَ ثُمَّ قلتُ هاتِ الآنَ ما عندَكَ فقالَ: حدَّثَنا ابنُ عباسٍ قالَ: قالَ رسولُ اللَّهِ صلى الله عليه وسلم: (صوموا لرؤيتِهِ وأَفطِروا لرؤيتِهِ، فإنْ حالَ بينَكُم وبينَهُ سحابٌ فكمّلوا العدةَ ثلاثينَ، ولا تستقبلوا الشهرَ استقبالاً).[11]

Simak ibn Harb reported: "I woke up one day feeling confused as to whether the day was [the last] day of Shaʿban or [the first of] Ramadan, so I started the day fasting. Then I went to ʿIkrimah and saw him eating bread and herbs. He said, 'Come and have lunch with me,' so I told him I was fasting. Thereupon he said, 'I swear by Allah that you must break your fast.' When I saw him swearing without making any exception, I went forward and asked for his proof. He said that Ibn ʿAbbas reported that the Prophet said, "Fast by sighting it [the new moon], and break your fast by sighting it. Complete thirty days if it is cloudy, and do not immediately fast before the month [of Ramadan] starts.'"

حدثنا مُحمَّدُ بنُ الصَّباحِ الْبَزَّازُ أخبرنا جَرِيرُ بنُ عَبْدِ الْحَمِيدِ الضَّبِّيُّ عنْ مَنْصُورٍ بن الْمُعْتَمِرِ عن رِبْعِيِّ بن حِراشٍ عن حُذَيْفَةَ، قال: قال رَسُولُ الله صلى الله عليه وسلم: (لَا تُقَدِّمُوا الشَّهْرَ حتى تَرَوْا الهِلَالَ أوْ تُكْمِلُوا الْعِدَّةَ ثُمَّ صُومُوا حتى تَرَوْا الهِلَالَ أوْ تُكْمِلُوا الْعِدَّةَ). قالَ أَبو دَاوُدَ: رَوَاهُ سُفْيَانُ وَغَيْرُهُ عن مَنْصُورٍ عن رَبْعِيٍّ عن رَجُلٍ من أصحابِ النَّبِيِّ صلى الله عليه وسلم لَمْ يُسَمِّ حُذَيْفَةَ.[12]

> On the authority of Hudhayfah, who reported that the Prophet said, "Do not fast a day or two before Ramadan until you see the new moon or complete thirty days [of Shaᶜban]. Then fast until you see the moon [of Shawwal] or complete thirty days." Abu Dawud adds: "This hadith has been narrated by Sufyan and others on the authority of Mansur, who reported it from Rabiᶜi from a Companion of the Prophet who was not called Hudhayfah."

The Prophet stressed naked-eye sighting because it was the only available mechanism for the first Muslims to achieve the required certainty, not because the act of actually sighting was the objective of fasting or of any other act of worship. If this objective can be achieved by a more authentic and precise method, as is the case today (viz., calculations), then replacing naked-eye sighting (which is a probable means of certainty) with a more accurate method based on categorical certainty cannot be considered a deviation from the prophetic command or the Shariᶜah's objectives.

Second, if naked-eye sighting were such an objective or a prerequisite without which fasting could not begin, then it would have been required even on 30 Shaᶜban. But nobody goes out to see the new moon on 30 Shaᶜban or 30 Ramadan, and no jurist ever required that they do so. Once certainty has been attained by completing thirty days (no Islamic month cannot go beyond thirty days), naked-eye sighting is no longer required and the new month is confirmed because it is known that the new moon must be above the horizon by 30th Shaᶜban.

Third, if it is said that sighting is not required on 30 Shaᶜban because the Prophet said "complete thirty days if it is cloudy," then I would argue that this prophetic statement means exactly what it says. For example, if the evening of 29 Shaᶜban was clear and yet the new moon was not seen, would not the hadith have required Muslims to see it on 30 Shaᶜban if it were a precondition for fasting? The phrase "if it is cloudy" cannot be applied to a clear sky. If naked-eye sighting were intrinsic to fasting, then it would have been required even on thirtieth day, especially if the new moon had not been seen, not because of some obscurities but because it was still below the horizon. However, if we agree that certainty is the precondition for fasting, then one can understand why naked-eye sighting is required on 29 Shaᶜban but not on 30 Shaᶜban.

Fourth, sighting is not a prerequisite for fasting even on 29 Shaᶜban. If it were, then no Muslim would be allowed to start Ramadan before sighting the new moon on 29 Shaᶜban. But Ibn ᶜUmar, along with ᶜA'ishah and Asma' bint Abi Bakr, used to start fasting the next day if it was cloudy on

29 Shaʿban and the new moon was obscured. They would not fast that day as a optional day of fasting, but as a mandatory day of Ramadan.

This was also the practice of many of the Successors. In fact, Ahmad ibn Hanbal built his jurisprudential school upon the actions of these Companions. This becomes important when we realize that Ibn ʿUmar was the original narrator of many of the hadiths found in the hadith compilations that require naked-eye sighting as a means of fasting as well as requiring Muslims not to fast until they see the moon. For instance, he reports:

> حدّثنا عبدُ اللّهِ بنُ مَسْلمةَ عن مالكٍ عن نافعٍ عن عبدِ اللّهِ بنِ عُمَر رضيَ اللّهُ عنهما: (أنَّ رسولَ اللّهِ صلى الله عليه وسلم ذَكَرَ رَمضان فقال: لا تَصوموا حتّى تَرَوُا الهلالَ، ولا تُفْطِرُوا حتّى تَرَوْه، فإن غُمَّ عليكم فاقدُروا له).[13]

> The Prophet mentioned Ramadan and said, "Do not fast until you see the moon and do not break the fast until you see it. Estimate it when it is cloudy."

> حدثنا عبد الله، حدثني أبي، حدثنا عبد الرحمن، حدثنا مالك، عن نافع، عن ابن عمر، عن النبي صلى الله عليه وسلم قال: (لا تصوموا حتى تَروا الهلال، ولا تفطروا حتى تَرَوْه، فانْ غُمَّ عليكم فاقْدُرُوا له).[14]

> The Prophet mentioned Ramadan and said, "Do not fast until you see the moon and do not break the fast until you see it. Estimate it when it is cloudy."

ʿAbd Allah ibn ʿUmar's Practice

As we see, Ibn ʿUmar would start Ramadan by counting the days of Shaʿban and without actually sighting the new moon if it was cloudy on 29 Shaʿban. His practice explains the true meanings of such hadiths and repudiates the second most important argument of the majority: that there is a cause and effect relationship between naked-eye sighting and fasting. Logically speaking, sighting in itself cannot be the sole reason for determining Ramadan or the act of fasting. It must be a means to achieve the goal of certainty. The following points must be kept in mind when discussing this question.

Ibn Daqiq al-ʿId, a Shafiʿi authority, states:

> وَلَيْسَ حَقِيقَةُ الرُّؤْيَةِ بِشَرْطٍ مِنْ اللُّزُومِ؛ لانَّ الاتِّفَاقَ عَلَى أَنَّ الْمَحْبُوسَ فِي الْمَطْمُورَةِ إذَا عُلِمَ بِإِكْمَالِ الْعِدَّةِ، أَوْ بِالاجْتِهَادِ بِالامَارَاتِ: أَنَّ الْيَوْمَ مِنْ رَمَضَانَ، وَجَبَ عَلَيْهِ الصَّوْمُ وَإِنْ لَمْ يَرَ الْهِلالَ. وَلا أَخْبَرَهُ مَنْ رَآهُ.[15]

> Actual sighting is not a condition for the requirement [of fasting the month of Ramadan]. There is agreement [among the jurists] that if someone were imprisoned in the basement and knew, either by completing thirty days or by diligence [estimation] in following the signs, that the month of Ramadan had started, then he would be required to start fasting even if he had neither sighted the moon himself nor was informed by someone who had actually sighted it.

Interestingly, Hamza Yusuf translates the above quote as follows:

> The reason for this is that the actual physical sighting is not what is legally binding, given that it is agreed upon that if someone was confined in a cell and knew by calculation that the [month] had run its course or by attempting to understand the signs that the day was indeed from Ramadan, then it would be incumbent upon him to fast, even if he did not see the crescent and no one informed him that it had been seen.[16]

In addition, he indirectly criticizes me for supposedly misunderstanding Ibn Daqiq's original intent:

> Sadly, in the papers I examined that used this quote to support calculation, the first half of his quote was omitted, so that his actual position was entirely misrepresented. One paper stated that the imam did not consider sighting to be a condition, when what he is saying is that *physical sighting is not legally binding in the case of one who cannot see the moon*, which is why he uses as an example the man in the cell who has no access to sighting the moon nor to news of anyone who did! In such circumstances, the imam says one can resort to calculating or ijtihad.[17]

His translation of Ibn Daqiq's statement (the original Arabic is given below) as "The reason for this is that the actual physical sighting *is not what is legally binding*" is confusing.

وَلَيْسَ حَقِيقَةُ الرُّؤْيَةِ بِشَرْطٍ مِنْ اللُّزُومِ.

Ibn Daqiq's words clearly state that "actual sighting is not the condition of requiring [fasting]." This supports the Qur'anic and hadith understanding that even those who do not actually see the new moon but are informed of Ramadan's arrival by others should fast, given they are not among those who are exempted (e.g., the sick, travelers, and expectant or nursing mothers). Indeed, requiring naked-eye sighting as a precondition for fasting would exclude from fasting all those who did not actually see the new moon. Such an understanding violates the long-standing Muslim scholarly consensus. Perhaps this is what Ibn Daqiq is stating in his example of a prisoner in the basement or a cell. Had the Shariʿah required naked-eye sighting as the only condition for starting or ending Ramadan, then that particular person, who would have no access to the outside world and no way to learn about Ramadan's arrival, would not have to fast. And yet the consensus requires such a person to fast based upon personal calculations and following the possible signs.

In consideration of this, what does his statement "One paper stated that the imam did not consider sighting to be a condition, when what he is saying is that *physical sighting is not legally binding in the case of one who cannot see the moon…*" mean? Can physical sighting be legally binding if one cannot see the moon?

Furthermore, in this particular paper I mentioned that Ibn Daqiq opposes the use of calculations to determine Ramadan and ʿId al-Fitr on either 29 Shaʿban or 29 Ramadan, unless there are obscurities. Like many other jurists, he maintains that only two legal causes (*sabab sharʿī*) require fasting: naked-eye sighting and completing thirty days. He differs from others in asserting that calculations can be used to confirm Ramadan if there are obscurities on 29 Shaʿban. This is a legal cause for depending upon calculations. After quoting Mutarrif ibn ʿAbd Allah ibn al-Shakhir and many Shafiʿi juristic authorities as advocating the use of calculations to confirm the month, Ibn Daqiq states:

وَالَّذِي أَقُولُ بِهِ: إِنَّ الْحِسَابَ لا يَجُوزُ أَنْ يُعْتَمَدَ عَلَيْهِ فِي الصَّوْمِ، لِمُفَارَقَةِ الْقَمَرِ لِلشَّمْسِ، عَلَى مَا يَرَاهُ الْمُنَجِّمُونَ، مِنْ تَقَدُّمِ الشَّهْرِ بِالْحِسَابِ عَلَى الشَّهْرِ بِالرُّؤْيَةِ بِيَوْمٍ أَوْ يَوْمَيْنِ. فَإِنَّ ذَلِكَ إِحْدَاثٌ لِسَبَبٍ لَمْ يُشَرِّعْهُ اللَّهُ تَعَالَى.[18]

My opinion on this matter is that the astronomical calculations based upon the separation of the moon from the sun cannot be depended upon in fasting because they push forward the month a day or two before the actual sighting. This is introducing a cause (*sabab*) that Allah has not prescribed.

On the other hand, he considers clouds or any other obscurity that hampers naked-eye sighting to be a lawful cause (*sabab sharʿī*) to determine fasting by calculation. The following is the actual text:

وَأَمَّا إِذَا دَلَّ الْحِسَابُ عَلَى أَنَّ الْهِلالَ قَدْ طَلَعَ مِنَ الافُقِ عَلَى وَجْهٍ يُرَى، لَوْلا وُجُودُ الْمَانِعِ كَالْغَيْمِ مِثْلا فَهَذَا يَقْتَضِي الْوُجُوبَ، لِوُجُودِ السَّبَبِ الشَّرْعِيِّ. وَلَيْسَ حَقِيقَةُ الرُّؤْيَةِ بِشَرْطٍ مِنَ اللُّزُومِ؛ لانَّ الاتِّفَاقَ عَلَى أَنَّ الْمَحْبُوسَ فِي الْمَطْمُورَةِ إِذَا عُلِمَ بِإِكْمَالِ الْعِدَّةِ، أَوْ بِالاجْتِهَادِ بِالامَارَاتِ: أَنَّ الْيَوْمَ مِنْ رَمَضَانَ، وَجَبَ عَلَيْهِ الصَّوْمُ وَإِنْ لَمْ يَرَ الْهِلالَ. وَلا أَخْبَرَهُ مَنْ رَآهُ.[19]

> If the calculations show that the new moon is born in the horizon to the extent that it could be seen were it not for obscurities such as clouds, then this makes it obligatory to fast owing to the presence of a lawful Islamic cause. The actual sighting is not a prerequisite of fasting. There is agreement [among the jurists] that if someone was imprisoned in the basement and knew, either by completing thirty days or by diligence in following the signs, that the month of Ramadan has started, then he would be required to observe the fast even if he had neither sighted the moon himself nor been informed by someone who had.

There is a possibility that he intended to state that naked-eye sighting in case of obscurities is not a condition for the obligation to fast. His choice of the given example, however, points more to the fact that naked-eye sighting is not the sole condition for this obligation, but no more than a means to the end of witnessing the month. The Qur'anic meanings of this phrase were analyzed in the previous section.

Sighting the moon is not a Legal Requirement

Saʿd al-Din Masʿud ibn ʿUmar al-Taftazani, a Hanafi, describes a consensus on the fact that naked-eye sighting is just a means, not the end in itself:

> أَنَّ قوله تعالى (فَمَنْ شَهِدَ مِنْكُمْ الشَّهْرَ) مَعْنَاهُ شَاهَدَ الشَّهْرَ فَالشُّهُودُ عِلَّة وَأَيْضًا قَوْلُهُ عليه الصلاة والسلام (صُومُوا لِرُؤْيَتِهِ) يَدُلُّ عَلَى ذَلِكَ إِذْ لَيْسَ الْمُرَادُ حَقِيقَةَ الرُّؤْيَةِ إِجْمَاعًا بَلْ مَا يَثْبُتُ بِهَا وَهُوَ شُهُودُ الشَّهْرِ.[20]

> The meaning of the Qur'anic verse "whosoever witnesses the month" refers to [a person] witnessing the month [being present in the month]. Consequently, witnessing the month is a cause or *ʿillah* [of fasting], and the prophetic statement "fast on sighting it" proves this. All Muslim jurists agree that the actual sighting is not meant [here, meaning that it is not the objective], but [that the objective is] what the sighting proves, and that is the witnessing of the arrival of the month.

Mustafa al-Zarqa summarizes this point as follows:

> وما دام من البديهات أن رؤية الهلال الجديد ليست فى ذاتها عبادة فى الإسلام، وإنما هي وسيلة لمعرفة الوقت، وكانت الوسيلة الوحيدة الممكنة فى أمة أمية لا تكتب ولا تحسب، وكانت أميتها هي العلة فى الأمر بالاعتماد على العين الباصرة، وذلك بنص الحديث النبوي مصدر الحكم.[21]

> It is an established fact that sighting the new moon in itself is not an act of Islamic worship. It is just a means to know the time. It was the only way available to a

> nation that did not know how to write or calculate. Its unlettered status was the sole reason for it to depend upon the naked eye. This is precisely what the text of the hadith, which is the original source of this ruling [fast on sighting], states.

In conclusion, the Prophet prescribed naked-eye sighting because it was the only method available to his illiterate community that could ensure certainty concerning the beginning or end of a given month.

THE MULTIPLE CONNOTATIONS OF *RA'Ā*

The verb *ra'ā*, *yarā* or "seeing" or "sighting" is generally used in the hadith quoted above in the sense of naked-eye sighting, although linguistically the verb is not confined to that. Maghraoui, however, states:

> The word "seeing" (*ru'yā*) in the second text must be interpreted in its literal (*ḥaqīqī*) sense in accordance with the basic rule in legal hermeneutics: a word must be taken in its literal and not metaphorical (*majāzī*) meaning unless it is impossible to understand it literally and there is a circumstantial reason (*qarīnah*) for assigning it a metaphorical meaning. *It is not impossible to understand* ru'yā *to mean seeing. The physical sighting of the moon was always the practical understanding of the word* ru'yā.[22]

The verb *ra'ā – yarā* has been used elsewhere in both the Qur'an and the hadiths where the meanings cannot denote naked-eye sighting but only pondering or certainty. The verb "seeing" and its many derivatives occur in the Qur'an 328 times. In a number of these verses, "seeing" is used in the context of pondering or ascertaining without resorting to naked-eye sighting. Examples of this use can be found in 2:243 and 2:46.

(ألم تر إلى الذين خرجوا من ديارهم وهم ألوف حذر الموت فقال لهم الله موتوا ثم أحياهم إن الله لذو فضل على الناس ولكن أكثر الناس لا يشكرون.) (البقرة: 243)

> Did you not turn your vision to those who abandoned their homes, though they numbered thousands, for fear of death? Allah said to them: "Die": then He restored them to life. For Allah is full of bounty to humankind, but most of them are ungrateful. (2:243)

(ألم تر إلى الملإ من بني إسرائيل من بعد موسى إذ قالوا لنبي لهم ابعث لنا ملكا نقاتل في سبيل الله.) (البقرة: 246)

> Have you not turned your vision to the Chiefs of the Children of Israel after [the time of] Moses? They said to a prophet [that was] among them: "Appoint for us a king, so that we may fight in the cause of Allah." (2:246)

The same meaning can be understood from 2:258, 3:23, 4:44, 4:49, 4:60, 96:9, 96:13, 107:1, and many more verses. Furthermore, numerous authentic narrations imply the same verb "seeing" with the above-mentioned connotation of "knowing" in relation to fasting Ramadan. In the following agreed-upon hadith, in which the Prophet deploys "seeing" for the night, the actual sighting of the night does not seem to be the objective, for specifying its timings or certainty is more important.

حدّثنا مُسدَّدٌ حدَّثنا عبدُ الواحِدِ حدَّثنا الشَّيْبَانِيُّ قال: سمعتُ عبدَ الله بنَ أبي أوفى رضيَ الله عنهُ قال: (سِرْنا معَ رسولِ الله صلى الله عليه وسلم وهوَ صائمٌ، فلما غَرَبتِ الشمسُ قال: انزل فاجدَحْ لنا، قال: يارسولَ الله لو أمسَيتَ؟ قال: انزِلْ فاجدَحْ لنا، قال: يارسولَ الله إنَّ عليكَ نهاراً، قال: انزِلْ فاجدَحْ لنا، فَنَزَلَ فجدَحَ، ثم قال: إذا رأَيتُم الليلَ أقْبلَ مِن ها هنا فقد أفطَرَ الصائمُ)، وأشارَ بإصبَعِه قِبَلَ المَشرِق.[23]

We traveled with the Prophet while he was fasting. At sunset, he said, "Dismount and mix the drink for us." He was asked, "What if we wait for the night?" He said, "Prepare the drink." He was told that it still appeared to be daylight. He again asked for the drink, which was prepared for him. Then he said, "When you see the night approaching from there (and he pointed toward the East), the fasting person should break the fast."

Muslim clearly reports that this journey was made during Ramadan.[24]

In these hadiths, the Prophet uses *ra'aytum* in connection with breaking the fast of Ramadan. If we take his statement "If you see the night coming from the East" literally, then we would have to go out every evening to actually see the night coming from the East in order to break our fast. Today nobody does this; all Muslims just follow the clock to determine the time of *ifṭār*. As the Muslims of the Prophet's time did not have many choices, they adopted the method that would provide the most certainty: looking toward the East and seeing the night approaching as a sign of sunset. Times have changed, however, and no contemporary jurist objects to using a clock. The same principle applies to determining the time of *saḥūr* and *imsāk:*

(وكلوا واشربوا حتى يتبين لكم الخيط الأبيض من الخيط الأسود من الفجر ثم أتموا الصيام إلى الليل...) (البقرة: 187)

And eat and drink until the white thread of dawn appears to you distinct from its black thread; then complete your fast till the night appears. (2:187)

Furthermore, for many centuries the Ummah followed the shadow cast by poles to determine the time for *ẓuhr* and *ʿaṣr*. The Prophet himself advised Muslims to do so. Today we use clocks to determine these times. I am not arguing here that the prayer timess are connected with the solar system whereas the time for Ramadan is connected with the lunar system and naked-eye sighting. Rather, I am arguing that although the Qur'an and the Sunnah require us to differentiate between a white thread and a black thread at dawn to begin our fast or to see the night approaching from the East, we now implement them only in the spirit, not in the letter, because following these specific commands literally has never been the real objective.

For example, the Law ensures that the sun has set and the night has approached before anybody breaks the fast. In the case of *imsāk*, whether dawn had arrived or not had to be ascertained. This objective, which has remained the same throughout Islamic history, can now be achieved by means of clocks because the Ummah does not object to their use in religious matters. These significant matters are, in turn, connected with the obligatory acts of fasting and praying. In other words, the Shariʿah's objectives are constants, whereas the means vary according to the circumstances. This is the Shariʿah's true spirit, for such flexiblity enables it to keep up with modern developments and progress.

Means versus Goals in Islamic Law: Discussion of the Constants and Variables

Some contemporary Muslims argue that the Shariʿah has also fixed the means to achieve Islamic goals. For instance, lawful (*ḥalāl*) earning is a means to achieve the goal of feeding one's family, for Muslims are not permitted to steal, cheat, or use some other unlawful means to achieve this goal. Likewise, the objective of fatherhood or generational continuity cannot be achieved by the unlawful means of adultery or fornication.

Abdur Rahman Ijaz asks: "Shouldn't all the three parts of this equation [be] as important as each other?"

Means - Methods - Achieve Goals

> If the goal is to feed the family, then the means and methods employed by the father are highly important…If [a] man wants to achieve fatherhood, are [not] the means and methods important, i.e. its due process! Should we accept your new paradigm [that] the achievement of fatherhood is important, then [the] method [itself].[25]

Feeding the family by stealing or cheating is *ḥarām* because Allah has clearly prohibited such practices. The only alternative or compromise permitted is if the person is facing a life-threatening situation. Becoming a father through unlawful means is also prohibited: "Do not even come close to adultery." Clearly, any means that are unlawful in and of themselves are prohibited.

The case of naked-eye sighting as a means to start or end Ramadan is quite different. The Qur'an prescribed "witnessing the month" as the cause of fasting. This cause can be understood in light of the prophetic reports demanding either naked-eye sighting or completing thirty days in the case of obscurities on the evening of the twenty-ninth day. The hadith about "estimation" in the case of cloudy weather clearly states another option as a means to determine Ramadan. The fact that there is no requirement for naked-eye sighting on the evening of the thirtieth day gives another authentic means to confirm or negate Ramadan. In sum, the Prophet gave us naked-eye sighting and several other options to achieve the desired goal of certainty.

The Qur'an mentions several kinds of means, among them:

(وَأَعِدُّوا لَهُم مَّا اسْتَطَعْتُم مِّن قُوَّةٍ وَمِن رِّبَاطِ الْخَيْلِ تُرْهِبُونَ بِهِ عَدْوَّ اللَّهِ وَعَدُوَّكُمْ وَءَاخَرِينَ مِن دُونِهِمْ لا تَعْلَمُونَهُمُ اللَّهُ يَعْلَمُهُمْ وَمَا تُنفِقُوا مِن شَيْءٍ فِي سَبِيلِ اللَّهِ يُوَفَّ إِلَيْكُمْ وَأَنتُمْ لا تُظْلَمُونَ.)
(الأنفال: 60)

Against them make ready your strength to the utmost of your power, including steeds of war, to strike terror into [the hearts of] the enemies of Allah and your enemies, and others besides, whom you may not know, but whom Allah does know. Whatever you spend in the cause of Allah shall be repaid to you, and you shall not be treated unjustly. (8:60)

Here Allah clearly mentions horses as a means of striking fear into the enemy's heart. In the following hadith, the Prophet explicitly asks for the use of arrows as a means to achieve the military might needed for self-defense and victory.

حدثنا أبو كريب، قال: ثنا سعيد بن شرحبيل، قال: ثنا ابن لهيعة، عن يزيد بن أبي حبيب، وعبد الكريم بن الحرث، عن أبي عليّ الهمدانيّ، أنه سمع عقبة بن عامر على المنبر يقول: قال الله: وأَعِدُّوا لَهُمْ ما اسْتَطَعْتُمْ مِنْ قُوَّةٍ وَمِنْ رِباطِ الخَيْلِ ألا وإني سمعت رسول الله صلى الله عليه وسلم يقول على المنبر (قالَ اللَّهُ: وأعِدُّوا لَهُمْ ما اسْتَطَعْتُمْ مِنْ قُوَّةٍ ألا أنَّ القُوَّةَ الرَّمْيُ ألا إنَّ القُوَّةَ الرَّمْيُ) ثَلاثا.[26]

> On the authority of Abu Kurayb ... he heard ʿUqbah ibn ʿAmir recite from the pulpit, "Against them make ready your strength to the utmost of your power, including steeds of war" and then state that he heard the Prophet say, "Certainly [militayr] might consists of archery; certainly [military] might is archery."

ʿIkrimah maintains that this verse clearly requires Muslims to acquire fortresses and mares as means to achieving this desired might.

حدثنا ابن وكيع، قال: ثنا أبي، عن سفيان، عن شعبة بن دينار، عن عكرمة، في قوله: وأَعِدُّوا لَهُمْ ما اسْتَطَعْتُمْ مِنْ قُوةٍ قال: الحصون. وَمِنْ رِباطِ الخَيْلِ قال: الإناث.[27]

In modern warfare, Muslims would look stupid and be unnable to survive an hour of modern military firepower if they followed the above-mentioned Qur'anic and prophetic means of military might literally. Although these specified means were suitable at that time, such is no longer the case. Therefore, adopting modern means of military might (e.g., tanks, missiles, and combat aircraft) to achieve the desired goals of securing justice and defending Muslims must be considered obligatory. Once again, the goals are constant but the means are variable.

Calculations as a Means to Attaining Certainty

Likewise, calculations are a means to achieve certainty. The prophetic tradition did not ask for its use for several reasons, such as the practice among the Jews and the pre-Islamic Arabs of fixing calendars by arbitrary intercalations, the Ummah's almost total illiteracy, the difficulty associated with making calculations, and the undeveloped nature of astronomy at that time. The seemingly most significant reason, which could also be deduced from the authentic prophetic reports, was the Muslims' almost total illiteracy – something that was also true of the majority of later generations of Muslims. The majority of the classical scholars were right in rejecting the use of calculation to determine Ramadan because its results were inaccurate and mostly computed by immature astrologers and magicians.

In contrast, modern calculations are made by fully qualified scientists and astronomers who base their knowledge on scientific observation and facts. The margin or possibility of error is almost zero. Such arguments of inaccuracy and magic that were used to prevent or prohibit the use of calculations were rejected by Islamic jurists even in classical times. What about twenty-first-century America, where astronomy has reached its climax? Abdillah bin Bayyah observes:

ففي هذا الزمان تطور هذا العلم تطوراً كبيراً، بحيث كما يقول البعض: إن الإبرة اذا ارسلت في الجو يمكن الاهتداء اليها، فكيف القمر الذي يكون قد ولد أو خرج من الاقتران، وخرج من شعاع الشمس.[28]

The science of [astronomy] has made tremendous progress in our time. Some even say that it can guide us to a needle thrown into space. What about the moon that is born or has moved out of conjunction and from the rays of the sun?

Zayn al-Din ibn Ibrahim ibn Nujaym, a Hanafi, argued in favor of the precise calculations since early times:

نَقَلَ فِي الامْدَادِ عَنْ شَرْحِ الْمَنْظُومَةِ لابْنِ الشِّحْنَةِ أَنَّ الْمُرَادَ بِالْكَاهِنِ وَالْعَرَّافِ فِي الْحَدِيثِ مَنْ يُخْبِرُ بِالْغَيْبِ أَوْ يَدَّعِي مَعْرِفَتَهُ فَمَا كَانَ هَذَا سَبِيلَهُ لا يَجُوزُ، وَيَكُونُ تَصْدِيقُهُ كُفْرًا أَمَّا أَمْرُ الاهِلَّةِ فَلَيْسَ مِنْ هَذَا الْقَبِيلِ بَلْ مُعْتَمَدُهُمْ فِيهِ الْحِسَابُ الْقَطْعِيُّ فَلَيْسَ مِنْ الاخْبَارِ عَنْ الْغَيْبِ أَوْ دَعْوَى مَعْرِفَتِهِ فِي شَيْءٍ أَلا تَرَى إِلَى قوله تعالى (وَقَدَّرَهُ مَنَازِلَ لِتَعْلَمُوا عَدَدَ السِّنِينَ وَالْحِسَابَ).[29]

Ibn al-Shikhnah says that the magician and astrologist mentioned in the hadith are the those who describe the Unseen or claim knowledge of the future. The statement of such a person will not be accepted, and approving these claims will constitute an act of disbelief. However, calculations of the [phases of the] moon have nothing to do with that. They are based upon precise data and are not fortune-telling or anything connected with the Unseen. Do not you see what Allah says in the Qur'an? "He prescribed for [the moon] computed stages so that you can learn about the numbers of years and the calculation."

Taqi al-Din ʿAli ibn ʿAbd al-Kafi al-Subki explains this hadith:

وَلا يَعْتَقِدُ أَنَّ الشَّرْعَ أَبْطَلَ الْعَمَلَ بِمَا يَقُولُهُ الْحِسَابُ مُطْلَقًا فَلَمْ يَأْتِ ذَلِكَ، وَكَيْفَ وَالْحِسَابُ مَعْمُولٌ بِهِ فِي الْفَرَائِضِ وَغَيْرِهَا، وَقَدْ ذُكِرَ فِي الْحَدِيثِ الْكِتَابَةُ وَالْحِسَابُ، وَلَيْسَتْ الْكِتَابَةُ مَنْهِيًّا عَنْهَا فَكَذَلِكَ الْحِسَابُ.[30]

It cannot be that the Shariʿah has categorically prohibited the usage of astronomical calculations. That is not so. How could this be when calculations are being used for the obligatory as well as other matters [of *dīn*]? The oft-quoted hadith mentions writing and calculations. When writing is not forbidden, how could astronomical calculations be [forbidden]?

Regarding the hardship argument, it must be noted that we now live in a "global village." In this age of instant communication, news is broadcast world-wide within seconds. Thus the argument of hardship leveled by al-Nawawi and others loses its ground. In reality it is the other way round, as

Yusuf al-Qaradawi rightly contends.[31] Muslims all over the globe and especially in the West suffer much hardship due to uncertainties connected with naked-eye sighting. Some of them wait until midnight just to start their *tarāwīḥ* prayers or decide about their ʿId prayers. Muslim employees and students also face many hardships. Therefore, depending upon naked-eye sighting instead of calculations has itself become the source of hardship.

It is also a fact that calculations and their usage in matters of *dīn* and *ʿibādah* are nothing new, for calculations have been used for quite some time to determine the time for the prayers, as well as for *saḥūr*, *ifṭār*, and also the direction of the qiblah. Throughout Islamic history, all jurists have not only accepted such practices but have also required Muslims to learn about them.

قَسَّمَ الْفُقَهَاءُ عِلْمَ النُّجُومِ إِلَى قِسْمَيْنِ: الاوَّلُ: حِسَابِيٌّ: وَهُوَ تَحْدِيدُ أَوَائِلِ الشُّهُورِ بِحِسَابِ سَيْرِ النُّجُومِ. وَيُسَمَّى مَنْ يُمَارِسُ ذَلِكَ الْمُنَجِّمَ بِالْحِسَابِ. وَلا خِلافَ بَيْنَ الْفُقَهَاءِ فِي جَوَازِ مُمَارَسَةِ التَّنْجِيمِ بِهَذَا الْمَعْنَى، وَتَعَلُّمِ مَا يُعْرَفُ بِمَوَاقِيتِ الصَّلاةِ وَالْقِبْلَةِ، بَلْ ذَهَبَ جُمْهُورُهُمْ إِلَى أَنَّ ذَلِكَ فَرْضُ كِفَايَةٍ. وَجَاءَ فِي حَاشِيَةِ ابْنِ عَابِدِينَ: وَالْحِسَابِيُّ حَقٌّ، وَقَدْ نَطَقَ بِهِ الْكِتَابُ فِي قَوْلِ الْحَقِّ تَبَارَكَ وَتَعَالَى: (الشَّمْسُ وَالْقَمَرُ بِحُسْبَانٍ). وَأَجَازَ الْفُقَهَاءُ الاعْتِمَادَ عَلَيْهِ فِي دُخُولِ أَوْقَاتِ الصَّلاةِ وَتَحْدِيدِ جِهَةِ الْقِبْلَةِ وَقَالُوا: إِنَّ حِسَابَ الاهِلَّةِ، وَالْخُسُوفِ وَالْكُسُوفِ قَطْعِيٌّ، فَاللَّهُ سُبْحَانَهُ وَتَعَالَى أَجْرَى حَرَكَاتِ الافْلاكِ وَانْتِقَالاتِ الْكَوَاكِبِ عَلَى نِظَامٍ وَاحِدٍ دَائِمٍ، وَكَذَلِكَ الْفُصُولُ الارْبَعَةُ. وَالْعَوَائِدُ إِذَا اسْتَمَرَّتْ أَفَادَتْ الْقَطْعَ، فَيَنْبَغِي الاعْتِمَادُ عَلَيْهِ فِي أَوْقَاتِ الصَّلاةِ وَنَحْوِهَا، وَفِي جِهَةِ الْقِبْلَةِ.[32]

Jurists have divided the knowledge of stars into two categories. First is the calculation of celestial bodies and their movements to determine the beginning of the months. The person who practices this kind of astronomy is called an astronomer. There is no disagreement among jurists that this exercise is permitted. It is permitted to learn this science in an effort to know the times of the prayers and the directions of the qiblah. Indeed, the majority of jurists (*jumhūr*) opine that this knowledge must be acquired by a number of Muslims at all times. Ibn ʿAbidin, in his *Hāshiyah*, says that "astronomical calculations are Islamically approved." This is exactly what the Qur'an says: "The sun and the moon follow meticulous calculations." Jurists have allowed the dependence on calculations concerning the times of the daily prayers as well as the directions of the qiblah. Astronomical calculations connected with new moons and lunar and solar eclipses are correct. Almighty God has fixed a system for the celestial bodies, and they always follow that system fully. The same principle applies to the four seasons. The aspects of nature that are continually repeating themselves are categorical in nature. Therefore, they should be relied upon in the matters of the prayer times as well as the direction of the qiblah.

Ahmad ibn Muhammad al-Hamwi stated the same opinion:

وَأَمَّا مُجَرَّدُ الْحِسَابِ مِثْلُ ظُهُورِ الْهِلالِ فِي الْيَوْمِ الْفُلانِيِّ وَوُقُوعِ الْخُسُوفِ اللَّيْلَةَ الْفُلانِيَّةَ فَإِنَّهَا أُمُورٌ حِسَابِيَّةٌ مَبْنِيَّةٌ عَلَى أَرْصَادٍ وَاقِعَةٍ فَلا تَدْخُلُ فِي نَهْيِ النَّبِيِّ صلى الله تعالى عليه وسلم، وَيُؤَيِّدُهُ مَا يُجَوِّزُونَهُ مِنْ تَعْلِيمِ قَدْرِ مَا تُعْلَمُ بِهِ مَوَاقِيتُ الصَّلاةِ وَالْقِبْلَةِ.[33]

> Calculations relating to new moons and eclipses are based upon actual realities and experiments. They do not come under the category of acts prohibited by the Prophet. This argument is substantiated by the fact that jurists have allowed the knowledge of calculations when it comes to knowing the times of the daily prayers and the direction of the qiblah.

How could it be that jurists who declare knowledge of calculations as obligatory (*farḍ kifāyah*) for determining the time of the obligatory prayers rule that it is totally un-Islamic in regard to fasting Ramadan? That is why Mustafa al-Zarqa is amazed that many contemporary conservative jurist adamantly reject calculations the latter instance even though they use them in the former – especially since the daily prayers are far more important and frequent than fasting Ramadan. Classical jurists were right to oppose calculations, for the methods available at their time had not reached the level of authenticity and certainty that we presently enjoy. How could they have based important acts of worship upon calculations that were not completely precise and accurate? Are we going to drag their opposition to calculations into our own time, a time in which the underlying reasons for their opposition is no longer valid? Obviously, if the cause is no longer present then the effect must also cease to exist.[34]

Hamza Yusuf and Astronomy

Hamza Yusuf argues that astronomy had already become developed during the early Muslim centuries and that

> the early Muslims were masters of observational astronomy and far superior in their observations than modern astronomers, who do almost all of their work in theoretical abstractions and mathematics, and are largely uninterested in actual physical observation of objects and events that are discernable to the unaided naked-eye, especially since the ancients have completed that work, and there is little left to be explored.[35]

In addition:

> Many modern Muslims, even those highly educated, believe there has been great advancement in observational astronomy. The telescope has undeniably enabled modern man to penetrate the heavens in unimaginable ways, and to understand

> elliptical cycles of the moons and planets as well as the development of calculus. However, naked-eye astronomy has changed very little. In fact, the reality remains that the observed motion of the sun, moon, planets, and stars is far less understood to the common man as well as most astronomers than to pre-modern people…[36]

He quotes al-Qarrafi as a scholar of astronomy and as one who believed that astronomy was decisive.[37]

I do not deny that Muslim scientists played a major role in developing many medieval sciences. But it has been proven that medieval astronomy was based upon some incorrect wrong principles, such as geo-centrism instead of solo-centrism. Al-Qarrafi is a good example of what I am arguing here, for he states that astronomical calculations made according to conjunctions are precise and decisive. At the same time, however, he mentions the names of the seven planets known at his time: Saturn, Jupiter, Mars, the sun, Venus, Mercury, and the moon. This is sufficient to refute the assertion that the observational astronomy used by the early Muslim scientists was far superior to modern astronomy, given that we now know that

> nine major planets are currently known. They are commonly divided into two groups: the inner planets (Mercury, Venus, Earth, and Mars) and the outer planets (Jupiter, Saturn, Uranus, and Neptune). The inner planets are small and are composed primarily of rock and iron. The outer planets are much larger and consist mainly of hydrogen, helium, and ice. Pluto does not belong to either group, and there is an ongoing debate as to whether Pluto should be categorized as a major planet.[38]

In addition, there are many satellites, asteroids, and comets, as well as interplanetary dust and gas, that also move around the sun. Al-Qarrafi's clarification that there are only seven moving planets tells the whole story.

The true context of what Hamza Yusuf has quoted from al-Qarrafi's[39] *Al-Furuq* is that the iman, in these statements, is responding to those who had criticized him as being inconsistent.[40] They contended that sighting the sun's movement in relation to the daily prayers was just as obligatory as sighting the moon for fasting Ramadan. The authentic prophetic hadiths require sighting in both cases. But why did al-Qarrafi differentiate between the two required sightings (allowing calculations to determine the prayer times, instead of sighting the sun's movement, but denying their use to determine Ramadan)? He answered by stating that the mandatory legal cause of confirming the prayer time was to attain certainty about the arrival of the specific prayer's actual time. It was perfectly acceptable to achieve that certainty by any means other than actual sighting of the sun's movements. On the other hand, naked-eye sighting of the new moon was required as a legal cause of Ramadan. The imam states:

(فَإِنْ قُلْت) هَذَا جُنُوحٌ مِنْك إِلَى أَنَّهُ لا بُدَّ مِنْ الرُّؤْيَةِ، وَأَنتَ قَدْ فَرَّقْت بَيْنَ الْبَابَيْنِ، وَمَيَّزْت بَيْنَ الْقَاعِدَتَيْنِ بِالرُّؤْيَةِ وَعَدَمِهَا، وَقُلْت السَّبَبُ فِي الاهِلَّةِ الرُّؤْيَةُ وَفِي أَوْقَاتِ الصَّلَوَاتِ تَحْقِيقُ الْوَقْتِ دُونَ رُؤْيَتِهِ فَحَيْثُ اشْتَرَطْت الرُّؤْيَةَ فَقَدْ أَبْطَلْت مَا ذَكَرْته مِنْ الْفَرْقِ.

قُلْت سُؤَالٌ حَسَنٌ (وَالْجَوَابُ عَنْهُ) أَنِّي لَمْ أَشْتَرِطْ الرُّؤْيَةَ فِي أَوْقَاتِ الصَّلَوَاتِ لَكِنِّي جَعَلْت عَدَمَ اطِّلاعِ الْحِسِّ عَلَى عَدَمِ الْفَجْرِ دَلِيلا عَلَى عَدَمِهِ وَأَنَّهُ فِي نَفْسِهِ لَمْ يَتَحَقَّقْ؛ لانَّ الرُّؤْيَةَ هِيَ السَّبَبُ وَنَظِيرُهُ فِي الاهِلَّةِ لَوْ كَانَتْ السَّمَاءُ مُصْحِيَةً وَالْجَمْعُ ثير وَلَمْ يُرَ الْهِلالَ جَعَلْت ذَلِكَ دَلِيلا عَلَى عَدَمِ خُلُوصِ الْهِلالِ مِنْ شُعَاعِ الشَّمْسِ.[41]

If you say that [by the example you presented] you are leaning toward the fact that sighting [the sun] is required [for prayer timings], you have already differentiated between the two scenarios and the two foundational rules by requiring actual sighting in one scenario [Ramadan] and not in the other [prayer timings]. You maintain that the cause [of fasting] in connection with the moons is actual sighting, while the cause of prayer timings is certainly about arrival of the prayer time without actual sighting [of the sun]. Therefore, when you require actual sighting regarding the moon, you [in reality] nullify the claimed difference between the two.

Al-Qarrafi seems to be stating here that naked-eye sighting is the legal cause of starting Ramadan on the evening of 29 Shaᶜban. It is impossible to prove that the Lawgiver has ever required naked-eye sighting on 30 Shaᶜban. In addition, there is no legal requirement for naked-eye sighting if the skies are cloudy. Therefore, naked-eye sighting may be accepted as a legally binding cause of starting to fast after 29 Shaᶜban. The cause and the effect always go hand in hand, but this is not true as regards naked-eye sighting and fasting Ramadan. Al-Shatibi has stated:

ما أثبت سبباً، فهو سبب أبداً لا يرتفع...[42]

The established [legal] cause always remains the cause and never changes.

DEFINING THE ISLAMIC LEGAL CAUSE

ᶜAbd al-Karim Zaydan defines the legal cause (*al-sabab al-sharᶜī*) as follows:

ما جعله الشارع معرفاً لحكم شرعي، بحيث يوجد هذا الحكم عند وجوده و ينعدم عند عدمه.[43]

The cause is what the Lawgiver has established in order to define the legal ruling in such a way that this specific ruling exists with its existence and disappears with its absence.

How could naked-eye sighting be the sole binding legal cause of fasting if Ramadan could begin by sighting, completion of thirty days, or mere estimation? It seems that many jurists who have insisted upon naked-eye sighting as the sole legal cause for fasting have confused *al-ḥukm al-taklīfī* and *al-ḥukm al-waḍʿī*. In Islamic jurisprudence, the former denotes a divinely commissioned ruling that requires Muslims either to act or not to act upon a demand, or it gives them a choice either to perform or not to perform an act. This kind of *ḥukm* includes almost all of the various categories of acts: mandatory, recommended, permitted, disliked, and prohibited.

وهو ما يقتضي طلب الفعل، أو الكف عنه، أو التخيير بين فعله وتركه، وهو يشمل الوجوب
والاستحباب والإباحة والكراهة والتحريم، ومن أمثلة ذلك وجوب الصلاة.[44]

(*Al-ḥukm al-taklīfī*) is a divine command that requires an action or prohibits it, or gives a choice between performing or abandoning it. It includes the mandatory, recommended, permitted, disliked, and prohibited acts. The example of (*al-ḥukm al-taklīfī*) is the requirement of the daily prayers.

On the other hand, the positional or positivistic divine command (*al-ḥukm al-waḍʿī*) is situational in nature. It identifies the cause, condition, or the reason for the prohibition of a ruling or a specific practice. Moreover, it is just a description of a specific ruling and not a demand from the Lawgiver to perform or abandon an act. For instance, noon is a cause of performing *ẓuhr* and theft results in a mandatory punishment. If there is no theft, there is no punishment. In other words, the existence of that particular *ḥukm waḍʿī* does not mean that the Lawgiver requires stealing.

الحكم الشرعي الوضعي: وهو ما يقتضي جعل شيء سبباً لشيء، أو شرطاً له، أو مانعاً منه، وهو
بالتالي ليس فيه أي طلب من المكلّف بفعل أو ترك، بل هو بيان من الشارع مثل: اعتبار زوال
الشمس سبباً لوجوب الصلاة، أو اعتبار السرقة سبباً لوجوب الحدّ.[45]

(*Al-ḥukm waḍʿī*) is what requires one thing as a cause or condition for something else or negates such a condition. Consequently, it does not contain any demand for an action or non action by a believer. It constitutes the Lawgiver's explanation (for a specific ruling), for instance (His) considering noon as the cause of requiring the (*ẓuhr*) prayer or considering theft as cause of capital punishment.

Therefore, the *ḥukm al-taklīfī* is "fasting Ramadan" and the *hukm al-waḍʿī* is "witnessing Ramadan." Naked-eye sighting in the past was just a means to determine the act of witnessing; it was neither the objective nor cause in itself. As seen earlier, the established legal causes do not change, for they are constants.

Al-Nawawi states that fasting Ramadan is connected with the timing of [the entrance, coming of] Ramadan.

ولا يجب صوم رمضان إلا بدخوله ويعلم دخوله برؤية ...[46]

Fasting Ramadan does not become obligatory until the month arrives, and the arrival is known by sighting the new moon.

Ibn ᶜAbd al-Barr has stated the point very neatly:

God says, "Whoever witnesses the month should fast" (2:185). He means, and God knows best, "Whoever among you knows, *with a knowledge that is certain*, that the month has indeed begun must fast it." Knowledge that is certain is [based on] either a clear and widespread sound sighting or the completion of thirty days of the previous month.[47]

From these classical quotations, Faysal Mowlawi draws the logical conclusion that the coming of Ramadan makes fasting obligatory and that witnessing the month ascertains that coming of Ramadan. Therefore, witnessing the month is the real cause and not the sighting, because the "witnessing" takes place both with and without sighting, as Al-Razi states:

أن شهود الشهر بماذا يحصل؟ فنقول: إما بالرؤية وإما بالسماع.[48]

How is witnessing the month accomplished? We say that it is achieved either by physical sighting or by hearing.

Thus sighting is just one of the means (*waṣīlah*) to determine Ramadan's arrival; it is not the legal cause (*sabab*) of fasting. Mowlawi, who differentiates between a cause (*sabab*) and a condition (*ᶜillah*), concludes:

ومعنى ذلك أنّ دخول رمضان هو سبب وجوب الصيام، وأنّ رؤية الهلال هي وسيلة العلم بدخوله.[49]

It means that the coming of the month of Ramadan is the cause of fasting Ramadan and that the sighting of the moon is a method of knowing that coming [of Ramadan].[50]

Furthermore, it is quite challenging to describe sighting the moon as the legal cause of fasting and then explain why the Lawgiver specified it as a cause only for 29 Shaᶜban and not for the following day, or why He gave the options of sighting, completion, or estimation. On the other hand, if we confirm that the objective of certainty about the Ramadan's arrival as the real reason for the prophetic insistence upon naked-eye sighting on 29 Shaʿban, then we can fully understand why He did not require it on the evening of 30 Shaʿban (to confirm the fasting of Ramadan). It is also quite important to understand this specific prophetic insistence within the context of the real challenges and confusion caused by arbitrary intercalations.

Rashid Rida, Mustafa al-Zarqa, Ahmad Shakir, and many others have highlighted the fact that the Lawgiver allowed naked-eye sighting only as a means to determine such fixed sacred times as Ramadan, not to worship the act of sighting in itself. The Prophet connected both naked-eye sighting and completion due to the wide-spread illiteracy, for the objective of his prophethood was to educate the Ummah, not to push it further into illiteracy.

Moreover, there is no ijmaʿ (agreement or consensus) among the jurists that naked-eye sighting is the only cause of fasting Ramadan, although many classical jurists emphasized it as the legal cause of confirming Ramadan. The real cause, as stated in the Qur'an, is "witnessing the month." This can be done through naked-eye sighting as well as other means of knowledge about Ramadan. Many jurists have stated this fact. For instance, Muhammad Ahmad Mustafa Ahmad Abi Zahrah states that the cause of fasting is the month of Ramadan.

(الشهر هو إمارة على وجوب الصوم).[51]

The month [of Ramadan] is the cause/sign of the obligation of fasting.

ʿAbd Allah ibn al-Judayʿ states that it is the coming of Ramadan.

(دخول الشهر لوجوب صوم رمضان) أخذاً من قوله تعالى (فمن شهد منكم الشهر فليصمه...) (سورة البقرة: 185).[52]

The Qur'anic verse "whoever witnessed the month should fast" makes the arrival of the month the cause of obligatory fasting during Ramadan.

ʿAbd al-Karim Zaydan also states that the legal cause of fasting is the coming of the month.

أمثلة السبب الشرعي (دلوك الشمس لوجوب الصلاة، وشهر رمضان لوجوب الصيام).[53]

Examples of legal cause are sunset [is the cause] for prayer obligation and the month of Ramadan for obligatory fasting.

That is why it is important that we understand al-Qarrafi's statements about naked-eye sighting of the moon as the sole legal cause of fasting and also sighting the sun in regard to the prayer times in light of the imam's dialogue with his opponents. It cannot be denied that he, like many other classical jurists, maintained that naked-eye sighting was the legal cause of fasting Ramadan. What is being denied here is that his verdict on this issue is relative to his milieu; it was not absolute that it must be applied forever. As seen above, he disagreed with the established authorities of almost all of the legal schools of thought by negating the unity of horizons and affirming their variety. Early jurists declared that the unity of horizons was the legal binding cause of the month of Ramadan. Modern jurists are following in his footsteps by disagreeing with his claim that naked-eye sighting is the sole legal cause of fasting. In fact, the sole legal cause of fasting is witnessing the month, and naked-eye sighting is just one of the means to achieve that goal. We should keep this point in mind when analyzing the classical juristic positions on naked-eye sighting.

CHAPTER 4

A BRIEF HISTORICAL EXPOSITION OF THE CLASSICAL DEBATE ABOUT SIGHTING THE MOON AND CALCULATIONS

With very few exceptions, classical Islamic scholars required naked-eye sighting as a precondition to confirm or negate Ramadan. They opposed calculations and rejected their use to determine both Ramadan and whether the new moon had been sighted. A detailed scrutiny of their discourse will show that they rejected and actually prohibited all calculation-based methods for any aspect of Ramadan. Their opposition to the imprecise astrologically computed calculations makes a lot of sense when understood within the context of intercalation, astrology, fortune telling, imprecision, hardship, and similar matters that directly affect faith and dogma.

It is also important to consider their cultural and political milieu when discussing their attitude toward naked-eye sighting. Many classical scholars, among them Ibn Taymiyyah, actively confronted the extreme allegorists such as the Isma^cilis, philosophers (e.g., Neoplatonists), and anti-traditionalist liberals (e.g., Mu^ctazilites). These so-called liberals attempted to intellectualize Islam and its texts at the expense of its original context, meanings, and intents. As a result, the traditionalists prohibited any variance between the actual texts and what they agreed to be the literal meanings.

However, this frequently promoted ijma^c (consensus) existed only among the jurists of a given *fiqhī* school, an ideological movement, or even only in the mind of a particular jurist without any accompanying support among others or within his/her own intellectual world. For instance, Ibn Taymiyyah and Ibn al-Qayyam insisted upon some quite literal understandings of such descriptive Qur'anic phrases as the "hand of Allah," "face of Allah," "eyes of Allah," and "chair and throne of Allah." They asserted that the linguistically proven literal meanings of "hand," "face," "eyes," and so on were the originally intended meanings of the Qur'an and Sunnah as long as any comparison between those belonging to Him and those belonging to His creatures is avoided. Both imams' selective insistence that "hand" denotes the "actual hand" (as long as Allah's hand is clearly understood to be different from the human hand and without *kayf* or asking how it looks), and their opposition to other possible metaphorical connotations of "hand" (e.g., support and protection), is a good way to elaborate the point under discussion. Ibn Taymiyyah, trying to avoid allegorizing the text, insisted upon the text's literal sense over the metaphorical sense. In fact, he ruled that the Muᶜtazilites were unbelivers as well as (indirectly) the later Ashᶜarites, among them al-Ghazali and many other experts in *tafsīr*, hadith, and fiqh.

This literally prone conservative ideological movement expressed extreme concern about interpretations that did not include the original linguistic meanings. They enforced the exact prophetic action as well as the Companions' actions and interpretations as the sole intended Sunnah and true explanation of several ambiguous verses. This group of ideologues was quite aware of the havoc that intellectualizing or liberalizing the sacred text could wreak on the Qur'an's original intent. They quoted this fact in their writings along with numerous examples, such as the role played by Philo of Alexandria in the Jewish understanding of the Hebrew Bible and his concept of "Logos" as the first emanated "Intellect."[1]

Ibn Taymiyyah also knew the role that this concept had played later in Christian theological thought. The early Church Fathers[2] borrowed Philo's concept and incorporated it into a fully fledged incarnational theology – the incarnation of God in the person of the historical Jesus of Nazareth – to such an extent that he was completely lost. In his *Jesus According to Paul*, Paul Furnish writes: "Paul focuses his attention neither on the teachings of Jesus nor on Jesus' Palestinian ministry. His attention is focused, rather, on Jesus the crucified Messiah and the risen Lord."[3] John Hick observes: "Paul fits Jesus into his own theology with little regard to the historical figure."[4]

This allegorical interpretation of the sacred text enabled St. Augustine of Hippo to categorize scriptural meanings into literal (apparent), allegorical, spiritual, and mystical. As a result, the New Testament was so mangled in the attempt to find spiritual and mystical meanings that the original content was lost.

Keeping in mind this comparative development of the concept of allegorical interpretations in Christianity and Judaism, Ibn Taymiyyah and many other traditionalists struggled to preserve the Qur'an's original intent and original form by emphasizing the literal meanings of such Qur'anic terms as the hand, face, and eyes of Allah.[5]

Al-Ghazali, al-Razi, and other theologians considered it inappropriate to describe such things in physical terms. Instead, they interpreted His "hand" as His power, dominance, and support, as in "I will give you a hand." This does not mean "I will cut off my hand and give it to you," but that "I will lend you support." This established metaphorical use has been linguistically and culturally approved. Many of these jurists/theologians refuted such phrases' literal meanings and maintained the principle of assigning the true understanding of the phrases to the knowledge of Allah (*al-tafwīḍ*). Al-Ghazali and others considered any insistence upon the literal meanings of such words to be sheer anthropomorphism, whereas Ibn Taymiyyah considered their metaphorical interpretations (*ta'wīl*) tantamount to disbelief.[6]

Traditionalist scholars had to fight many battles on multiple fronts to preserve what they believed to be the texts' precise nature. Convinced that the Ummah's degeneration into petty factions and separate states, as well as its loss of political might, was linked directly to its departure from the sublime Sunnah. A good example was the Mongols' utter destruction of the Muslim world; Ibn Taymiyyah witnessed this period of political and social degradation. What these scholars struggled for was nothing short of implementing the pristine faith and restoring the Prophet's lost *sunan* (traditions) to the best of their understanding.

On the other hand, many academicians, theologians, jurists, and scholars right through to modern times have criticized them for some very narrow and extreme interpretations of the Islamic texts, as already seen in the case of M. Zahid al-Kawthari. Among contemporary academicians, Abdul Aziz Sachedina blames this selective insistence upon some meanings at the expense of other accepted meanings, and then claiming divine propriety for these human interpretations to the exclusion of others, for Islamic fundamentalism and all other acts of Muslim violence.[7] Khaled Abul Fadl writes extensively about the possible dangers of this selective explanation.[8]

I will not go as far as to hold this traditional insistence on the letter of the Qur'an and Sunnah (and to use that literal process at the expense of the texts' spirit) as responsible for such modern political problems as terrorism. I fully recognize the sincerity that goes into past and contemporary scholarly attempts to preserve the Sunnah in its pristine form. However, I would like to state that this kind of narrow selection, if carried to extreme limits, can often challenge the spirit as well the holistic meanings of the Qur'anic and the prophetic texts. It goes beyond the established objectives (*maqāṣid*) of the Shari'ah and results in endless hardship for many sincere Muslims. The Qur'anic principle of "ease" (*taysīr*) is meant to make people's lives easy as long as the Shariʿah's pristine spirit is not violated – and that is exactly what Allah demands of us in 2:185:

> Ramadan is the [month] in which was sent down the Qur'an, as a guide to humankind, also clear [signs] for guidance and judgment [between right and wrong]. So

> whosoever witnesses the month among you should fast in it [spend it in fasting], but if anyone is ill, or on a journey, the prescribed period [should be made up] by days later. Allah intends ease for you; He does not want to put you in difficulties. [He wants you] to complete the prescribed period, and to glorify Him in that which He has guided you; and perhaps you shall be grateful. (2:185)

The following part of it can be particularly helpful to our discussion:

> Allah intends ease for you; He does not want to put you in difficulties. [He wants you] to complete the prescribed period, and to glorify Him in that which He has guided you…

In light of this, let's glance at the classical position on determining the Islamic months by calculations as opposed to naked-eye sighting. This long-standing debate seems to have begun even before the jurisprudential schools of thought appeared. Mutarrif ibn ʿAbd Allah ibn al-Shakhir, one of the Successors, is reported to have been the first person to use calculations to determine Ramadan when the sky was cloudy on 29 Shaʿban. The first known imam, Ahmad Abu Hanifah, died in 150 AH. By the time of Imam Malink the debate over calculations seemed to have developed into a full-scale *fiqhī* discussion. Imam Malik opposed Mutarrif's position on using calculations to determine any aspect of Ramadan or fasting. In addition, he is reported to be the first proponent of the argument that the hadith of *ikmāl* is descriptive of the hadith on "calculation," as will be seen below.

It seems that the issue of *ikmāl* was not yet developed during the era of Companions like Ibn ʿUmar and others, who fasted on 30 Shaʿban even if it had been cloudy on the previous day. Abu Bakr and ʿUmar, as well as ʿA'ishah and Asma' (Abu Bakr's daughters), would have never intentionally violated the prophetic ruling on completing thirty days in the case of obscurities on 29 Shaʿban. They fasted on the thirtieth day if the horizon was obscured on the previous day.

It seems that Ibn ʿUmar, the original narrator of the prophetic hadith requiring naked-eye sighting and completion in the case of cloudy weather, followed neither of these prophetic patterns in certain situations. For example, he would ask for naked-eye sighting on 29 Shaʿban and fast the next day if it was seen, fast after 30 Shaʿban if the moon had not been sighted on the previous day, and also fast the next day if the new moon was not sighted on the evening of 29 Shaʿban due to obscurities. In this case, he would fast without actually sighting the new moon and without completing thirty days of Shaʿban.

Abu Muhammad ibn Hazm al-Zahiri and many others have highlighted his practice and its implied conflict with the reported hadiths. We shall discuss this topic in detail when dealing with the weakness of the *ikmāl* argument. Here, I just want to make the point that the Companions seem to have had different opinions on this matter. Ibn ʿAbbas, Abu Hurayrah, Ammar ibn Yasir, and others agreed that thirty days must be completed in the absence of naked-eye sighting on 29 Shaʿban due to obscurities. But others, such as Ibn ʿUmar, opinioned that Muslims should fast the next day if the new moon was not seen beceause of obscurities.

Ahmad ibn Hanbal and many traditional scholars maintained that Muslims are required to fast 30 Shaʿban if nobody went out on the evening of 29 Shaʿban to sight the moon or the claim of sighting was made by untrustworthy individuals whose witness was rejected by the Muslim ruler. According to him, such a day is known as the "day of doubt." He also declared fasting on 30 Shaʿban a mandatory day of Ramadan in the absence of naked-eye sighting owing to obscurities on 29 Shaʿban.

عَنْ أَحْمَدَ أَنَّهُ خَصَّ يَوْمَ الشَّكِّ بِمَا إِذَا تَقَاعَدَ النَّاس عَنْ رُؤْيَةِ الْهِلالِ أَوْ شَهِدَ بِرُؤْيَتِهِ مَنْ لا يَقْبَلُ الْحَاكِمُ شَهَادَتَهُ، فَأَمَّا إِذَا حَالَ دُونَ مَنْظَرِهِ شَيْء فَلا يُسَمَّى شَكًّا.[9]

To Ahmad the "day of doubt" is confined only to the day when the people did not go out to sight the new moon or it was witnessed by those whose witness is not accepted by the ruler. It is not a "day of doubt" if the new moon was not sighted due to obscurities.

If the prophetic command to "complete thirty days if it is cloudy" been categorically decisive and legally mandatory (*wājib*), then none of the Companions (e.g., ʿUmar and Ibn ʿUmar) and prominent Successors (e.g., Mutarrif, Mujahid, and Ta'us), and imams like Ahmad ibn Hanbal would have fasted in the case of obscurity on the 29 Shaʿban, which would contradict the prophetic command to complete thirty days. Ibn Qudamah explains this point as follows:

وَقَدْ فَسَّرَهُ ابْنُ عُمَرَ بِفِعْلِهِ، وَهُوَ رَاوِيه، وَأَعْلَمُ بِمَعْنَاهُ، فَيَجِبُ الرُّجُوعُ إِلَى تَفْسِيرِه.[10]

Ibn ʿUmar explained [the hadith of sighting and estimation] with his own practice. It is [important because] he was its [original] narrator and the most knowledgeable about its [true] meanings. Therefore, it is obligatory to turn to his explanation [of this hadith].

The two divergent opinions of Imam Malik and Imam Ahmad converged against the use of calculations because the astrologists claimed access to the Unseen (e.g., the stars) and their calculations and assertions were inaccurate. The dogmatic implications of these astrological assertions were disastrous for the general public's faith and ʿ*aqīdah*. Moreover, since these imprecise calculations were not widely available, most of the classical jurists rejected them and banned their use to determine or negate Ramadan. Successors like Mutarrif ibn ʿAbd Allah and jurists like Ibn Surayj took an exceptional stand – they permitted calculations in the case of obscurities or for those who undertook them. These individuals maintained that the prophetic command requiring the completion of thirty days was actually addressed to the general public, and that the command to use "estimation or calculation" was aimed at the educated sector.

وَعَنْ مُطَرِّفٍ أَيْضًا أَنَّ الْعَارِفَ بِالْحِسَابِ يَعْمَلُ بِهِ فِي نَفْسِهِ. أَمَّا ابْنُ سُرَيْجٍ فَاعْتَبَرَ قَوْلَهُ صلى الله عليه وسلم: (فَاقْدُرُوا لَهُ): خِطَابًا لِمَنْ خَصَّهُ اللَّهُ تَعَالَى بِعِلْمِ الْحِسَابِ.[11]

Mutarrif also maintained that the one who knows calculations should personally act upon them. Ibn Surayj, on the other hand, maintained that the prophetic command "calculate for it" is addressed to the ones whom Almighty Allah has bestowed with knowledge of calculations.

The implications of Mutarrif and Ibn Surayj's positions were far reaching, and they were criticized by the majority of classical jurists.

Later, Taj al-Din al-Subki, another Shafiʿi, went one step further by not only accepting the use of precise calculations as an authentic means to determine Ramadan (in the case of obscurities), but by also rejecting the statements of witnesses who claimed to have sighted the new moon lest the calculations prove otherwise. He allowed the use of calculations to negate a stated naked-eye sighting if they proved that the new moon was not yet on the horizon. This big step forward earned him the criticism of the majority of classical scholars. The classical books of almost all *fiqhī* schools are filled with polemics against him for taking such a progressive position.

Al-Subki expressed his pain and frustration in the following words:

قَدْ يَحْصُلُ لِبَعْضِ الاغْمَارِ وَالْجُهَّالِ تَوَقُّفٌ فِيمَا قُلْنَاهُ وَيَسْتَنْكِرُ الرُّجُوعَ إِلَى الْحِسَابِ جُمْلَةً وَتَفْصِيلا وَيَجْمُدُ عَلَى أَنَّ كُلَّ مَا شَهِدَ بِهِ شَاهِدَانِ يَثْبُتُ، وَمَنْ كَانَ كَذَلِكَ لا خِطَابَ مَعَهُ وَنَحْنُ إِنَّمَا نَتَكَلَّمُ مَعَ مَنْ لَهُ أَدْنَى تَبَصُّرٍ وَالْجَاهِلُ لا كَلامَ مَعَهُ.[12]

Some recklessly ignorant [person] may hesitate to accept what we have stated. He might regard it as abhorrent to resort to calculations in part or in whole, and may be stuck with the idea that whatever is witnessed by two persons is proven. No conversation can take place with such a rigid person. We are talking to those who at least enjoy the [knowledge of] basic logic. We cannot talk to the ignorant.

Al-Qaradawi wonders what his opinion would have been on the question of naked-eye sighting if he were living in our modern times with all of its scientific developments in astronomy.

فكيف لو عاش السبكى إلى عصرنا هذا ورأى من تقدم علم الفلك ... كما أشرنا.[13]

In 1939, the famous hadith authority Ahmad M. Shakir had the courage to go one step further: he maintained that modern astronomical calculations were certain enough to ascertain Ramadan's arrival and end without necessarily resorting to naked-eye sighting. Having devoted his entire life to hadith literature and detailed study of the hadith sciences, he concluded – and was strongly criticized for doing so – that confirming the month of Ramadan, based upon the new moon's birth, was the closest position to that of the prophetic hadiths.

ولقد أرى قولى هذا أعدل الأقوال، وأقربها إلى الفقه السليم، وإلى الفهم الصحيح للأحاديث الواردة فى هذا الباب.[14]

I consider my position [statement] to be the most equitable [righteous] of [all] the positions. This position is the closest to sound understanding and to the true meanings of all the hadiths narrated on this subject [calculation and moon sighting].

In 2006, the Fiqh Council of North America, following his lead, accepted astronomical calculations in confirming the Islamic months and to establish a predetermined calendar based upon them. The European Fiqh Council adopted his position in May 2007. Both councils have been reprimanded for following the Sunnah of the Jews and abandoning that of the Prophet. So the saga continues.

CLASSICAL SCHOLARSHIP AND NAKED-EYE
SIGHTING AS THE CAUSE OF RAMADAN

In view of this brief historical exposition, we now turn to the classical texts themselves. Al-Jassas states:

قَالَ أَبُو بَكْرٍ: قَوْلُ رَسُولِ اللَّهِ صلى الله عليه وسلم: "صُومُوا لِرُؤْيَتِهِ" مُوَافِقٌ لِقوله تعالى: (يَسْأَلُونَكَ عَنْ الْأهِلّةِ قُلْ هِيَ مَوَاقِيتُ لِلنَّاسِ وَالْحَجِّ) وَاتَّفَقَ الْمُسْلِمُونَ عَلَى مَعْنَى الآية وَالْخَبَرِ فِي اعْتِبَارِ رُؤْيَة الْهِلالِ فِي إيجَابِ صَوْمِ رَمَضَانَ ، فَدَلَّ ذَلِكَ عَلَى أَنَّ رُؤْيَةَ الْهِلالِ هِيَ شُهُودُ الشَّهْرِ.[15]

The statement of the Prophet "Fast by sighting it" is in line with the Qur'anic verse that says, "They ask you about the new moons. Say: 'They are timings for people and for hajj.'" The Muslims have a consensus that the verse and the hadith mean considering the sighting of the crescent moon in requiring the fasting of Ramadan. It leads to [the fact] that the sighting of the new moon is what is termed "witnessing the month."

He concludes that naked-eye sighting is the only method prescribed by the Prophet for confirming Ramadan. If it cannot be determined by actual sighting on 29 Shaᶜban owing to unfavorable conditions (e.g., cloudy weather), then completing thirty days is required, for that is the original rule.

وَقَوْلِهِ صلى الله عليه وسلم (صُومُوا لِرُؤْيَتِهِ وَأَفْطِرُوا لِرُؤْيَتِهِ ، فَإِنْ غُمَّ عَلَيْكُمْ فَعُدُّوا ثَلاثِينَ) هُوَ أَصْلٌ فِي اعْتِبَارِ الشَّهْرِ ثَلاثِينَ ، إلا أَنْ يُرَى قَبْلَ ذَلِكَ الْهِلالُ ، فَإِنَّ كُلَّ شَهْرٍ غُمَّ عَلَيْنَا هِلالُهُ فَعَلَيْنَا أَنْ نَعُدَّهُ ثَلاثِينَ. هَذَا فِي سَائِرِ الشُّهُورِ الَّتِي يَتَعَلَّقُ بِهَا الاحْكَامُ، وَإِنَّمَا يَصِيرُ إِلَى أَقَلِّ مِنْ ثَلاثِينَ بِرُؤْيَةِ الْهِلالِ.[16]

In light of the prophetic hadith, the original rule is that the month consists of thirty days unless the new moon is sighted before that. We must count thirty days for every month when we are unable to see the moon owing to cloudy weather. This rule applies to all the months connected with Islamic rituals. Only the physical sighting of the new moon will make the month less than thirty days.

He also asserts that Muslim jurists have reached consensus to reject calculations in confirming or negating Ramadan.

فَالْقَائِلُ بِاعْتِبَارِ مَنَازِلِ الْقَمَرِ وَحِسَابِ الْمُنَجِّمِينَ خَارِجٌ عَنْ حُكْمِ الشَّرِيعَةِ. وَلَيْسَ هَذَا الْقَوْلُ مِمَّا يَسُوغُ الاجْتِهَادُ فِيهِ ، لِدَلالَةِ الْكِتَابِ وَنَصِّ السُّنَّةِ وَإِجْمَاعِ الْفُقَهَاءِ بِخلافِه.[17]

One who believes in the phases of the moon and the calculations of the astrologers is against the Shariᶜah. This is not the area of ijtihad, for the Qur'an, the Sunnah, and the consensus of the jurists are against it.

Badr al-Din al-ᶜAyni summarizes the classical juristic majority's opinion:

لا يصح اعتقاد رمضان إلّا برؤية فاشية أو شهادة عادلة، أو إكمال شعبان ثلاثين يوما، وعلى هذا مذهب جمهور فقهاء الأمصار بالحجاز والعراق والشام والمغرب، منهم مالك والشافعي والأوزاعي والثوري وأبو حنيفة وأصحابه وعامة أهل الحديث.[18]

The [month of] Ramadan can be based only on a widespread public sighting [of the new moon] or a trustworthy witnessing or the completion of thirty days of Shaᶜban. The majority of the jurists in the [major] cities of al-Hijaz, al-Iraq, al-Sham, and al-Maghrib maintain that. They include Malik, al-Shafiᶜi, al-Awzaᶜi, al-Thawri, Abu Hanifah and his disciples, and most of the *Ahl al-Ḥadīth*.

The reason for starting the months with actual sighting, according to al-Jassas, is to begin the acts of worship with certainty instead of basing them upon mere probabilities.

وَهَذَا قَوْلُ النَّبِيِّ صلى الله عليه وسلم: (صُومُوا لِرُؤْيَتِهِ وَأَفْطِرُوا لِرُؤْيَتِهِ، فَإِنْ غُمَّ عَلَيْكُمْ فَأَكْمِلُوا عِدَّةَ شَعْبَانَ ثَلاثِينَ). فَفَرَضَ عَلَيْنَا عِنْدَ غُمَّةِ الْهِلالِ إِكْمَالُ عِدَّةِ شَعْبَانَ ثَلاثِينَ يَوْمًا، وَإِكْمَالَ عِدَّةِ رَمَضَانَ ثَلاثِينَ يَوْمًا عِنْدَ غُمَّةِ هِلالِ شَوَّالٍ، حَتَّى يَدْخُلَ فِي الْعِبَادَةِ بِيَقِينٍ، وَيَخْرُجَ عَنْهَا بِيَقِينٍ. وَكَذَلِكَ ثَبَتَ عَنْ النَّبِيِّ صلى الله عليه وسلم مُصَرَّحًا بِهِ أَنَّهُ قَالَ: (لا تَصُومُوا حَتَّى تَرَوْا الْهِلالَ، وَلا تُفْطِرُوا حَتَّى تَرَوْهُ). وَقَدْ رَوَى التِّرْمِذِيُّ عَنْ أَبِي هُرَيْرَةَ عَنْ النَّبِيِّ صلى الله عليه وسلم أَنَّهُ قَالَ: (احْصُوا هِلالَ شَعْبَانَ لِرَمَضَانَ).[19]

This is what the Prophet says, "Begin fasting by sighting it and stop fasting by sighting it. If it is cloudy, then complete by counting thirty days of Shaʿban." He has required us to count thirty days of Shaʿban when it is cloudy and also count thirty days of Ramadan when it is cloudy [on 29 Ramadan] before starting the month of Shawwal. This is required so that we can start the acts of worship based upon certainty and stop the acts of worship based upon certainty. This is what the Prophet has manifestly commanded by another authentic saying, "Do not fast until you see the new moon and do not break the fast until you see the new moon." Al-Tirmidhi has narrated on the authority of Abu Hurayrah that the Prophet said, "Count the moon of Shaʿban to determine Ramadan."

Al-Jassas represents the view of the majority of classical jurists. The official Hanafi, Maliki, Shafiʿi, and Hanbali position is that calculations are not the authentic way to determining the months, for these must be confirmed either by naked-eye sighting or completion. In the following pages, we will see how these classical scholars argue in favor of this established position.

Ahmad ibn Muhammad al-Hamawi states:

الشَّرْطُ عِنْدَنَا فِي وُجُوبِ الصَّوْمِ وَالافْطَارِ رُؤْيَةُ الْهِلالِ وَلا يُؤْخَذُ بِقَوْلِ الْمُنَجِّمِينَ. وَفِي التَّهْذِيبِ عَلَى مَذْهَبِ الشَّافِعِيِّ رحمه الله: لا يَجُوزُ تَقْلِيدُ الْمُنَجِّمِ فِي حِسَابِهِ لا فِي الصَّوْمِ وَلا فِي الافْطَارِ.[20]

For us [Hanafis], the condition for the fast and breaking the fast is the sighting of the crescent [moon]; the calculation of the astrologists cannot be followed in this matter. In *Al-Tahdhīb*, according to the Shafiʿi school, it is also stated that astrological calculations cannot be trusted either for beginning or ending the month of fasting.

Muhammad ibn ʿAbd Allah al-Kharshi presents the Maliki position:

الصَّوْمَ يَثْبُتُ بِمَا تَقَدَّمَ لا بِقَوْلِ مُنَجِّم فَلا يَثْبُتُ بِهِ لا فِي حَقِّ غَيْرِهِ وَلا فِي حَقِّهِ هُوَ؛ لانَّ صَاحِبَ الشَّرْعِ حَصَرَ الثُّبوتَ فِي: الرُّؤْيَةِ، أَوْ الشَّهَادَةِ، أَوْ إِكْمَالِ الْعَدَدِ فَلَمْ يُخْبِرْ بِزِيَادَةٍ عَلَى ذَلِكَ فَإِذَا قَالَ الْمُنَجِّمُ مَثَلا: الشَّهْرُ نَاقِصٌ أَوْ زَائِدٌ لَمْ يُلْتَفَتْ إِلَى قَوْلِهِ، وَلا إِلَى حِسَابِهِ، وَقَعَ فِي الْقَلْبِ صِدْقُهُ أَمْ لا.[21]

Fasting cannot be observed by following the statement of an astrologer. Neither the astrologer nor anyone else can fast based upon that, because the Prophet has confined the fasting solely to the sighting of the witnesses or completing thirty days. No other method is prescribed. Therefore, no attention should be paid to the statement/calculations of the astrologer regarding the month, whether one believes in the precision of his calculations or not.

Muhammad ibn Ahmad al-Dasuqi (Maliki), who elaborates the same view, adds that Imam Malik is of the opinion that thirty days should be completed for all the months when it is cloudy and the possibility of sighting the moon is non-existent.[22] Imam Malik himself is reported to have said:

وَقَدْ رَوَى ابْنُ نَافِعٍ عَنْ مَالِكٍ فِي الْمَزنِيَّةِ فِي الامَامِ لا يَصُومُ لِرُؤْيَةِ الْهِلالِ وَلا يُفْطِرُ لِرُؤْيَتِهِ ، وَإِنَّمَا يَصُومُ وَيُفْطِرُ عَلَى الْحِسَابِ أَنَّهُ لا يُقْتَدَى بِهِ وَلا يُتَّبَعُ.[23]

Ibn Nafiʿa reported in *Al-Mazniyyah* that Malik was of the opinion that in fasting if an imam followed calculations instead of moon sighting then he should not be followed [or obeyed in daily prayers].

Abu al-Walid argues that one should compensate for the days one has fasted based upon the calculations and not upon sighting or completion.

قَالَ الْقَاضِي أَبو الْوَلِيدِ رضي الله عنه فَإِنْ فَعَلَ ذَلِكَ أَحَدٌ فَالَّذِي عِنْدِي أَنَّهُ لا يُعْتَدُّ بِمَا صَامَ مِنْهُ عَلَى الْحِسَابِ وَيَرْجِعُ إِلَى الرُّؤْيَةِ وَاكَمَالِ الْعَدَدِ فَإِنْ اقْتَضَى ذَلِكَ قَضَاءَ شَيْءٍ مِنْ صَوْمِهِ قَضَاهُ.[24]

Al-Qadi Abu al-Walid stated that (if someone fasted based upon calculations) then his fasting is not accepted as valid. He must resort to the sighting or completion method even if it required making up for some missed days.

THE MUSLIM UMMAH IS UNLETTERED

The Prophet said:

حَدَّثَنَا آدَمُ حَدَّثَنَا شُعْبَةُ حَدَّثَنَا الاسْوَدُ بْنُ قَيْسٍ حَدَّثَنَا سَعِيدُ بْنُ عَمْرٍو أَنَّهُ سَمِعَ ابْنَ عُمَرَ رَضِيَ اللَّهُ عَنْهُمَا عَنْ النَّبِيِّ صَلَّى اللَّهُ عَلَيْهِ وَسَلَّمَ أَنَّهُ قَالَ إِنَّا أُمَّةٌ أُمِّيَّةٌ لا نَكْتُبُ وَلا نَحْسُبُ الشَّهْرُ هَكَذَا وَهَكَذَا يَعْنِي مَرَّةً تِسْعَةً وَعِشْرِينَ وَمَرَّةً ثَلاثِينَ.[25]

We are an unlettered people; we neither write nor calculate. The month is like that and like that, meaning sometimes twenty-nine days and sometimes thirty days.

Muslim reports that the Prophet, while describing the month, folded his thumb the third time:

قَالَ سَمِعْتُ سَعِيدَ بْنَ عَمْرِو بْنِ سَعِيدٍ أَنَّهُ سَمِعَ ابْنَ عُمَرَ رَضِيَ اللَّهُ عَنْهُمَا يُحَدِّثُ عَنْ النَّبِيِّ صَلَّى اللَّهُ عَلَيْهِ وَسَلَّمَ قَالَ إِنَّا أُمَّةٌ أُمِّيَّةٌ لا نَكْتُبُ وَلا نَحْسُبُ الشَّهْرُ هَكَذَا وَهَكَذَا وَهَكَذَا وَعَقَدَ الابْهَامَ فِي الثَّالِثَةِ وَالشَّهْرُ هَكَذَا وَهَكَذَا وَهَكَذَا يَعْنِي تَمَامَ ثَلاثِينَ.[26]

We are an unlettered people; we neither read nor calculate. The month is like that and like that, and he folded his thumb the thirdtime, and the month is like this and this and this, meaning thirty days.

Based upon the above reports, Shihab al-Din ibn Ahmad al-Ramli (Shafiʿi) argues:

أَنَّ الشَّارِعَ لَمْ يَعْتَمِدْ الْحِسَابَ بَلْ أَلْغَاهُ بِالْكُلِّيَّةِ بِقَوْلِهِ نَحْنُ أُمَّةٌ أُمِّيَّةٌ لا نَكْتُبُ وَلا نَحْسِبُ الشَّهْرَ هَكَذَا وَهَكَذَا وَقَالَ ابْنُ دَقِيقِ الْعِيدُ الْحِسَابُ لا يَجُوزُ الاعْتِمَادُ عَلَيْهِ فِي الصِّيَامِ.[27]

The Prophet did not depend upon calculations at all but negated it by his statement "We are an unlettered nation, we neither write nor calculate … Ibn Daqiq al-ʿId stated that calculations cannot be the source of confirming the fasting [of Ramadan]."

Al-Nawawi, in *Al-Majmūʿ*, also quotes this hadith and gives almost the same reasons for the rejection of calculations:

وَمَنْ قَالَ بِحِسَابِ الْمَنَازِلِ فَقَوْلُهُ مَرْدُودٌ بِقَوْلِهِ صلى الله عليه وسلم فِي الصَّحِيحَيْنِ (إِنَّا أُمَّةٌ أُمِّيَّةٌ لا نَحْسِبُ وَلا نَكْتُبُ ، الشَّهْرُ هَكَذَا، وَهَكَذَا) الْحَدِيثَ قَالُوا: وَلأَنَّ النَّاسَ لَوْ كُلِّفُوا بِذَلِكَ ضَاقَ عَلَيْهِمْ؛ لأَنَّهُ لا يَعْرِفُ الْحِسَابَ إلا أَفْرَادٌ مِنْ النَّاسِ فِي الْبُلْدَانِ الْكِبَارِ، فَالصَّوَابُ مَا قَالَهُ الْجُمْهُورُ، وَمَا سِوَاهُ فَاسِدٌ مَرْدُودٌ بِصَرَائِحِ الاحَادِيثِ.[28]

The statement of those who talk about the moon phases is rejected by the prophetic report authenticated by (Bukhari and Muslim) that "we are unlettered preople; we neither write nor calculate. The month is like this and like this." It will cause people hardship if they are required to follow the calculations, for they are known only to a few people living mostly in big cities. Therefore, the majority position is the right position and whatever else is there is rejected by the authentic sayings of the Prophet.

Ibn Hajar explains the meaning of the hadith "we neither write nor calculate" in the following words:

و قوله: (لا نكتب ولا نحسب) تفسير لكونهم كذلك، وقيل للعرب أميون لأن الكتابة كانت فيهم عزيزة، قال الله تعالى (هو الذي بعث في الأميين رسولا منهم) ولا يرد على ذلك أنه كان فيهم من يكتب ويحسب لأن الكتابة كانت فيهم قليلة نادرة، والمراد بالحساب هنا حساب النجوم وتسييرها، ولم يكونوا يعرفون من ذلك أيضا إلا النـــزر اليسير، فعلق الحكم بالصوم وغيره بالرؤية لرفع الحرج عنهم في معاناة حساب التسيير واستمر الحكم في الصوم ولو حدث بعدهم من يعرف ذلك، بل ظاهر السياق يشعر بنفي تعليق الحكم بالحساب أصلا، ويوضحه قوله في الحديث الماضي "فإن غم عليكم فأكملوا العدة ثلاثين" ولم يقل فسلوا أهل الحساب، والحكمة فيه كون العدد عند الإغماء يستوي فيه المكلفون فيرتفع الاختلاف والنزاع عنهم.[29]

The hadith is a description of how they actually were [a reflection of their reality that they were unlettered]. The Arabs were called illiterate because writing skills were very rare among them. Allah has said, "It is He who has sent among the unlettered a messenger from among themselves." This fact cannot be refuted by [the assertion] that among them there were some [individuals] who could write and calculate. The reason is that writing [skills] were very rare among them. And *al-ḥisāb* here refers to the calculation of the stars [celestial bodies] and their movement [in their orbits]. They also had only a very little knowledge of it [calculation]. Consequently, he [the Prophet] connected the ruling on fasting and other things with the sighting so as to remove from them the difficulty of having to struggle with calculating the movements of the celestial bodies. This ruling would continue even if, later on, someone knew that [how to calculate]. However, the apparent context [of the hadith] gives the sense of not connecting the ruling with calculations in the first place and this is explained by his (the Prophet's) statement in the hadith previously discussed: "Complete by counting thirty if it is cloudy" and he did not say "Ask the astronomers." The wisdom behind this is that counting thirty [days] is easy for everyone (all those who are commissioned to fast). Therefore, it will protect people from argument and discord.

A Rather Awkward Interpretation of Ibn Hajar's Quotation

Hamza Yusuf translates the above statement in a slightly different way:

Indeed, Ibn Hajar and others understood the hadith "We are an unlettered community – we neither read nor calculate" to mean something entirely different. They did not interpret the Prophet's preface as an operative cause but rather as a descriptive statement, an important and necessary distinction in jurisprudence. Ibn Hajar provides the following explanation of the hadith:

"Calculate," here, refers to astronomy and to the orbits of the planets because only a handful of them knew such things at that time. Thus the Prophet has made the legal obligation of fasting contingent upon actual sighting in order to remove any burdens from his community, i.e., of having to struggle with computations of celestial orbits. This ruling continues even should later people be able to do that. Indeed, the apparent meaning of the hadith rejects any association of calculation with the legal ruling....

Nowhere did he say, "If it is obscured then ask the people of calculation."[30]

He goes on to conclude:

Ibn Hajar recognizes that only a small number of people knew much about astronomy at the time, which is not dissimilar to our current situation, given the vast numbers of illiterate Muslims alive today. But there were, indeed, among the first generation of Muslims some who knew how to calculate astronomical phenomena given that some were capable of producing an intercalated lunisolar calendar. Moreover, Ibn Hajar understood that the ruling was a permanent one and not, as some have said, one that is contingent upon the innumeracy of his community, and thus falsely concluded that if some people learned such things later, they could switch to determining their months by calculation.[31]

Some sentences of Imam Ibn Hajar in the above translation are manipulated, apparently so that they will have some specific implications far removed from Ibn Hajar's own intent. For example, the sentence: و لم يكونوا يعرفون من ذلك أيضا إلا النـــزر اليسير is translated as "*because only a handful of them knew such things at that time.*" This specific sentence is italicized to emphasize that "only a small number of people *knew much about astronomy at the time*, which is not dissimilar to our current situation, given the vast numbers of illiterate Muslims alive today. But there were, *indeed, among the first generation of Muslims some who knew how to calculate astronomical phenomena given that some were capable of producing an intercalated lunisolar calendar.*"

Frankly, the above translation of this sentence is faulty and the conclusions drawn are fanciful. I have yet to see any proof in the books of hadith, *sīrah*, and Islamic history that any of the Prophet's disciples knew much about calculations (used to intercalate the lunar calendar in order to synchronize it with the solar calendar) in general and calculations in particular. Al-Shatibi confirms that astronomy was not among the Arabs' sciences.

لأن ذلك لم يكن من معهود العرب ولا من علومها.[32]

Because it (astronomy) was neither commonplace nor a science of the Arabs.

It is true that the pre-Islamic Arabs used an intercalated calendar for hajj and other purposes. There is no proof that any Companions played any part in this intercalation. The prophetic description that "we are an unlettered people" supports this historical fact. It is very likely that these

Arabs sought help from their Jewish friends to fix their calendars. There is ample historical evidence for their close affinity with the Jews of Madinah and some other places, which continued even after the Prophet's migration there. Furthermore, fixing the calendar by intercalations did not require much sophistication as regards calculations. The average length of the lunar month had been known to the Greeks and Babylonians centuries before the Arabs. The rabbis used this basic astronomical knowledge to multiply and add extra days so that the lunar year could follow a solar year. The procedure needed more mathematical than astronomical calculations.

Ibn Hajar here denies that astronomy was quite developed during the time of the Companions and also that only a few of them knew much about it.

There is another flaw in the above translation of Ibn Hajar's text.

بل ظاهر السياق يشعر بنفي تعليق الحكم بالحساب أصلا، ويوضحه قوله في الحديث الماضي "فإن غم عليكم فأكملوا العدة ثلاثين" و لم يقل فسلوا أهل الحساب.

> But the apparent context (of the hadith) gives a sense not to connect the ruling with calculations in the first place. This is explained by his (the Prophet's) statement in the previously discussed hadith, "Complete counting thirty if it is cloudy." He did not say "Ask the astronomers."

Hamza Yusuf's rendition of the phrase بل ظاهر السياق يشعر as "indeed, the apparent meaning of the hadith" is also somewhat inaccurate. There is a big difference between بل ظاهر السياق "but the apparent context" and saying "indeed, the apparent meaning..." I am not asserting here that Ibn Hajar approved of calculations to confirm Ramadan, because he did not. Again, the possible reasons for his disapproval are also discernable from his own statement: he emphasized the difficulties of computing calculations and argued elsewhere in his book that they were the work of astrologers who were devils among humans (لأهم شياطين الإنس) and that their calculations were nothing but conjecture and guesswork. Consequently, they were not suitable to serve as the foundation of a legal ruling (*al-ḥukm al-sharʿī*). He contended that the hadith's apparent context also led to this understanding.

"The Month is Twenty-Nine": Misunderstanding the prophetic Hadith

Hamza Yusuf also asserts that the Prophet did not use common numerals but rather his hands to describe the lunar month.

> What I find profoundly interesting is Qadi Abu Bakr's point that the Prophet (pbuh) could very well have used the words "twenty-nine and thirty" when indicating the number of days possible in a lunar month. Had he done so, those he was speaking to would have understood him, as he was wont to state numbers on many other occasions as reported in sound hadith; he used high numbers, such as one million (*alfu alf*); he also used twenty-seven, twenty-five, and five; and he used the number ninety-nine in the sound hadith, "God has ninety-nine Names, one hundred less one; whoever enumerates them will enter Paradise." However, he chose not to state any numbers when showing the number of days in a lunar month, as if to deter people from thinking about enumeration specifically when it comes to determining the lunar months. Hence, instead of saying the words "twenty-nine and thirty," the Prophet (pbuh) actually used his blessed hands, showing with his fingers how many days are possible in the month, as if to emphasize using the most basic and fundamental human ability of sight. It is as though he were saying, "Look, see, perceive with your eyes the month, even upon my hands." This insistence upon sighting the moon illustrates so well "the sense in Islam that it is the immediate surrounding conditions, rather than any theoretical ones, that reflect the Divine will of God in its relation to men, and that it is these which should determine the sacred acts."[33]

This is a flawed understanding of the prophetic hadith, because the Prophet did use the number "twenty-nine" when describing a lunar month in multiple authentic hadiths. This is reported by al-Bukhari, Muslim, Abu Dawud, al-Tirmidhi, al-Nisa'i, Ahmad ibn Hanbal, and almost every book of hadith. I do not know how Hamza Yusuf could miss such an obvious fact. Of course the Prophet always made things easy for his disciples and, indeed, for his entire Ummah. While he did use his hands to illustrate the possible numbers of days in a lunar month, he did not avoid using numerals like twenty-nine to describe the lunar month.

Muslim reports the following:

حَدَّثَنِي هَارُونُ بْنُ عَبْدِ اللَّهِ وَحَجَّاجُ بْنُ الشَّاعِرِ قَالا حَدَّثَنَا حَجَّاجُ بْنُ مُحَمَّدٍ قَالَ قَالَ ابْنُ جُرَيْجٍ أَخْبَرَنِي أَبُو الزُّبَيْرِ أَنَّهُ سَمِعَ جَابِرَ بْنَ عَبْدِ اللَّهِ رَضِيَ اللَّهُ عَنْهُمَا يَقُولا اعْتَزَلَ النَّبِيُّ صَلَّى اللَّهُ عَلَيْهِ وَسَلَّمَ نِسَاءَهُ شَهْرًا فَخَرَجَ إِلَيْنَا صَبَاحَ تِسْعٍ وَعِشْرِينَ فَقَالَ بَعْضُ الْقَوْمِ يَا رَسُولَ اللَّهِ إِنَّمَا أَصْبَحْنَا لِتِسْعٍ وَعِشْرِينَ فَقَالَ النَّبِيُّ صَلَّى اللَّهُ عَلَيْهِ وَسَلَّمَ إِنَّ الشَّهْرَ يَكُونُ تِسْعًا وَعِشْرِينَ ثُمَّ طَبَّقَ النَّبِيُّ صَلَّى اللَّهُ عَلَيْهِ وَسَلَّمَ بِيَدَيْهِ ثَلاثًا مَرَّتَيْنِ بِأَصَابِعِ يَدَيْهِ كُلِّهَا وَالثَّالِثَةَ بِتِسْعٍ مِنْهَا.[34]

> Jabir ibn ʿAbd Allah narrates that the Prophet separated himself from his wives for a month. He came out to us on the morning of the twenty-ninth. Some of the people indicated to him that it was the morning of the twenty-ninth. The Prophet said, "Certainly, the month consists of twenty-nine [days]. Then the Prophet folded his both hands three times, twice folding [and opening] all of his fingers and the third time [opening] only nine.

Al-Bukhari reports on Anas ibn Malik's authority that the Prophet said that the "month certainly consisted of twenty-nine days."[35] Al-Tirmidhi categorized these specific prophetic hadiths (in which the Prophet categorically stated that the month consisted of twenty-nine days) as good and authentic:

قَالَ أَبو عِيسَى هَذَا حَدِيثٌ حَسَنٌ صَحِيحٌ

Abu ʿIsa stated that this report is *ḥasan ṣaḥīḥ*."[36]

If someone were to argue that this prophetic description of the lunar month with the "common numerals that mathematicians use" did not refer to Ramadan but to another lunar month, it would not matter, because a lunar month is still a month. I further state that there are numerous hadiths in which the Prophet referred to Ramadan and said that the month consisted of twenty-nine days. For instance, Abu Dawud reports:

حَدَّثَنَا سُلَيْمَانُ بْنُ دَاوُدَ الْعَتَكِيُّ حَدَّثَنَا حَمَّادٌ حَدَّثَنَا أَيُّوبُ عَنْ نَافِعٍ عَنْ ابْنِ عُمَرَ قَالَ قَالَ رَسُولُ اللَّهِ صَلَّى اللَّهُ عَلَيْهِ وَسَلَّمَ الشَّهْرُ تِسْعٌ وَعِشْرُونَ فَلا تَصُومُوا حَتَّى تَرَوْهُ وَلا تُفْطِرُوا حَتَّى تَرَوْهُ فَإِنْ غُمَّ عَلَيْكُمْ فَاقْدُرُوا لَهُ ثَلاثِينَ.[37]

Ibn ʿUmar reported that the Prophet said, "The month is twenty-nine [days]. Therefore, do not fast until you see it [the moon] and do not break the fast until you see it. Count thirty days if it is obscured from you.

Imam Muslim also narrates the same with a little variation.

و حَدَّثَنَا يَحْيَى بْنُ يَحْيَى وَيَحْيَى بْنُ أَيُّوبَ وَقُتَيْبَةُ بْنُ سَعِيدٍ وَابْنُ حُجْرٍ قَالَ يَحْيَى بْنُ يَحْيَى أَخْبَرَنَا و قَالَ الاخَرُونَ حَدَّثَنَا إِسْمَعِيلُ وَهُوَ ابْنُ جَعْفَرٍ عَنْ عَبْدِ اللَّهِ بْنِ دِينَارٍ أَنَّهُ سَمِعَ ابْنَ عُمَرَ رَضِيَ اللَّهُ عَنْهُمَا قَالَ قَالَ رَسُولُ اللَّهِ الشَّهْرُ تِسْعٌ وَعِشْرُونَ لَيْلَةً لا تَصُومُوا حَتَّى تَرَوْهُ وَلا تُفْطِرُوا حَتَّى تَرَوْهُ إِلا أَنْ يُغَمَّ عَلَيْكُمْ فَإِنْ غُمَّ عَلَيْكُمْ فَاقْدِرُوا لَهُ.[38]

Ibn ʿUmar reported that the Prophet said, "The month consists of twenty-nine nights. Therefore, do not fast until you see it [the moon] and do not break the fast until you see it, unless it is cloudy. Count it if it is obscured from you.

Imam al-Nasa'i reports:

أَخْبَرَنَا مُحَمَّدُ بْنُ الْمُثَنَّى قَالَ حَدَّثَنَا عَبْدُ الرَّحْمَنِ عَنْ سُفْيَانَ عَنْ الاسْوَدِ بْنِ قَيْسٍ عَنْ سَعِيدِ بْنِ عَمْرٍو عَنْ ابْنِ عُمَرَ عَنْ النَّبِيِّ صَلَّى اللَّهُ عَلَيْهِ وَسَلَّمَ قَالَ إِنَّا أُمَّةٌ أُمِّيَّةٌ لا نَكْتُبُ وَلا نَحْسُبُ الشَّهْرُ هَكَذَا وَهَكَذَا وَهَكَذَا ثَلاثًا حَتَّى ذَكَرَ تِسْعًا وَعِشْرِينَ.[39]

Ibn ʿUmar reported that the Prophet said, "Certainly, we are an unlettered people; we neither write nor calculate. The month is like that and like that and like that" (three times), until he mentioned twenty-nine.

Calculations are Connected with Magic and Astrology

One of the main reasons for these scholars' rejection of calculations is the close connection between astronomy and magic, which the Prophet has forbidden. For instance, Ibn Hajar strictly prohibits the use of calculation by quoting the prophetic sayings that warn Muslims about the evils of astrology, such as "Anyone learning any part of astrology is learning a part of magic." ʿUmar was quoted as saying, "Learn from astrology whatever part is helpful in guiding you over the land and ocean and then stop." Therefore, any part of astrology other than the directional symbols and signs is, according to Ibn Hajar, un-Islamic.[40]

Ibn Taymiyyah, a strong opponent of using astronomical calculations to confirm or negate Islamic months, argued that they could never lead to a reliable method of finding the crescent moon, and that he, like al-Jassas, also claimed that the scholars agreed on this matter.

اتَّفَقَ عُلَمَاءُ الشَّرِيعَةِ الأَعْلامُ عَلَى تَحْرِيمِ الْعَمَلِ بِذَلِكَ فِي الْهِلالِ . وَاتَّفَقَ أَهْلُ الْحِسَابِ الْعُقَلاء عَلَى أَنَّ مَعْرِفَةَ ظُهُورِ الْهِلالِ لا يُضْبَطُ بِالْحِسَابِ ضَبْطًا تَامًّا قَطُّ ؛ وَلِذَلِكَ لَمْ يَتَكَلَّمْ فِيهِ حُذَّاقُ الْحِسَابِ ؛ بَلْ أَنْكَرُوهُ ؛ وَإِنَّمَا تَكَلَّمَ فِيهِ قَوْمٌ مِنْ مُتَأَخِّرِيهِمْ تَقْرِيبًا وَذَلِكَ ضَلالٌ عَنْ دِينِ اللَّهِ وَتَغْيِيرٌ لَهُ شَبِيهٌ بِضَلالِ الْيَهُودِ.[41]

Mainstream scholars of the Shariʿah agree that using calculations in determining the new moon is forbidden. Wise astronomers also agree that there is no way to authentically determine the crescent moon from calculations. That is why expert astronomers do not use, but reject, calculation. Only a group from the later generations, out of ignorance, have used it. This is basically changing the *dīn* of Allah by misleading people and following the misguidance of the Jews in this matter.

Here, Ibn Taymiyyah seems to be referring to the Jewish Rabbinical Council's decision to adopt astronomical calculations as the authentic means of confirming the Jewish lunar months. Elsewhere, he expressed his opposition to such a practice:

ولا ريب انه ثبت بالسنة الصحيحة واتفاق الصحابة انه لا يجوز الاعتماد على حساب النجوم كما ثبت عنه فى الصحيحين انه قال انا امة امية لا نكتب ولا نحسب صوموا لرؤيته وأفطروا

لرؤيته والمعتمد على الحساب فى الهلال كما انه ضال فى الشريعة مبتدع فى الدين فهو مخطىء فى العقل وعلم الحساب.[42]

Undoubtedly calculations are rejected by the Sunnah as well as the consensus of the Companions, for the authentic hadith says … Therefore, the person who depends upon calculation is a misguided innovator, not only mistaken in matters of the Shariʿah but also in matters of logic and astrology.

Calculations are Inaccurate

Ibn Taymiyyah also argued that astronomical knowledge was misleading and a forbidden act in and of itself, because its disadvantages outweighed its advantages. He quoted several prophetic narrations to denounce astrology.[43] He also substantiated his point by a practical encounter that he had had with the so-called astronomers of his time, from which he concluded that the methodology of astronomical calculation was based purely upon falsehood and cheating.

وَهَكَذَا الْمُنَجِّمُونَ، حَتَّى أَنِّي لَمَّا خَاطَبْتُهُمْ بِدِمَشْق وَحَضَرَ عِنْدِي رُؤَسَاؤُهُمْ، وَبَيَّنْتُ فَسَادَ صِنَاعَتِهِمْ بِالادِلَّةِ الْعَقْلِيَّةِ الَّتِي يَعْتَرِفُونَ بِصِحَّتِهَا، قَالَ لِي رَئِيسٌ مِنْهُمْ: وَاللَّهِ إِنَّا نَكْذِبُ مِائَةَ كِذْبَةٍ حَتَّى نَصْدُقَ فِي كَلِمَةٍ.[44]

This is how the astrologers are! I, by logical arguments, proved the wrong nature of their profession when I debated with their chiefs in Damascus. One of them told me that, by God, we concoct a hundred lies so as to be able to produce one truth.

He further contended:

وَالأَدِلَّةُ الدَّالَّةُ عَلَى فَسَادِ هَذِهِ الصِّنَاعَةِ وَتَحْرِيمِهَا كَثِيرَةٌ، لَيْسَ هَذَا مَوْضِعَهَا، وَقَدْ ثَبَتَ فِي صَحِيحِ مُسْلِمٍ، عَنْ النَّبِيِّ صلى الله عليه وسلم أَنَّهُ قَالَ (مَنْ أَتَى عَرَّافًا فَسَأَلَهُ عَنْ شَيْءٍ لَمْ يَقْبَلْ اللَّهُ لَهُ صَلاةً أَرْبَعِينَ يَوْمًا). وَالْعَرَّافُ، قَدْ قِيلَ إِنَّهُ اسْمٌ عَامٌّ لِلْكَاهِنِ وَالْمُنَجِّمِ وَالرَّمَّالِ وَنَحْوِهِمْ.[45]

The arguments against this profession and its prohibition are too many. This is not a place to go into the details of that. It is sufficient to quote what Muslim narrated from the Prophet, "If someone asks an astrologer (*ʿarrāf*) about something [unseen], Allah will not accept his prayers for forty days. The term (*ʿarrāf*) denotes the magician, the astrologer, and the others."

This group of scholars suggests several punishments for those who use astronomy and calculations. For instance, Muhammad ibn Ahmad al-Alish points out that nobody, either the astrologer himself or anyone else, should fast according to calculations. It is also forbidden to approve of an astrologer. Such a person should be given the death penalty without any opportunity for repentance if he openly propagates the belief that the stars directly influence human destiny. He should be treated as an apostate if he conceals his beliefs but argues indirectly about the impact of stars on human life. He should be asked to repent and, if he refuses to do so, should be put to death. He will be a sinful believer if he takes the stars as signs indicating events in the world, yet does believe that the actual power lies with Allah, not with the stars.[46]

In the view of Ibn Rushd, astronomers must be disciplined.[47] Abu al-Sadat al-Mubarak ibn Muhammad al-Jazri calls them human devils, for they base their calculations on mere conjecture, hunches, and anticipation. He also quotes the hadith mentioned above, which connects the knowledge of astrology with the knowledge of magic.

لأنهم شياطين الإنس وقد جاء في بعض الأحاديث من اقتبس بابا من علم النجوم لغير ما ذكر الله فقد اقتبس شعبة من السحر.[48]

Because they are devils of the human race. Some prophetic reports stated that "whosoever acquired any chapter of astrology other than what Allah permitted in reality acquired a portion of magic."

A SUMMARY OF THE CLASSICAL MAJORITY'S ARGUMENTS AGAINST CALCULATION

This group's main arguments against the use of calculation as a valid source of determining the Islamic months can be summarized as follows:

1: As regards confirming or negating the Islamic months, especially that of Ramadan, the Shariʿah requires naked-eye sighting for only this can guarantee certainty. Physical sighting, in their view, seems to be the goal instead of the means. By sighting, they mean naked-eye sighting. They further assert the existence of a consensus among all classical scholars that naked-eye sighting or the completion of thirty days is the only way of confirmation. They reiterate that the prophetic narrations calling for estimation or calculation in the case of cloudy weather must be understood in the light of the narrations that require the completion of thirty days. That is what they believe the consensus to be. Ibn Taymiyyah defined the consensus as:

أن تجتمع علماء المسلمين على حكم من الاحكام وإذا ثبت إجماع الأمة على حكم من الأحكام
لم يكن لأحد أن يخرج عن اجماعهم فان الأمة لا تجتمع على ضلالة.[49]

Consensus occurs when Muslim scholarship agrees upon a ruling of one of the Islamic rules. No one is permitted to oppose this consensus because the Ummah does not agree upon something inherently wrong.

He also argued:

والتحقيق أن الاجماع المعلوم يكفر مخالفه كما يكفر مخالف النص بتركه.[50]

The reality is that one who contradicts an established consensus in fact commits an act of disbelief. It is just like rejecting an established religious text.

2: Calculations are hypothetical in nature and mere conjectures. They can never lead us to an authentic method of determining the beginning or end of the Islamic lunar months. Such classical scholars as Ibn Taymiyyah and al-Jassas also seem to have claimed agreement among the *jumhūr* about rejecting them altogether.

3: Following calculations causes hardship for the general public, for its knowledge is confined to a few individuals mostly living in big cities, and so on, as argued by al-Nawawi.

4: Dealing with calculations and the movements of celestial bodies is the profession of magicians and fortune tellers, aspects of divination forbidden by the Shariʿah. The Prophet forbade this by saying:

حَدَّثَنَا أَبُو بَكْرِ بْنُ أَبِي شَيْبَةَ وَمُسَدَّدٌ الْمَعْنَى قَالا حَدَّثَنَا يَحْيَى عَنْ عُبَيْدِ اللَّهِ بْنِ الاخْنَسِ عَنْ الْوَلِيدِ
بْنِ عَبْدِ اللَّهِ عَنْ يُوسُفَ بْنِ مَاهَكَ عَنْ ابْنِ عَبَّاسٍ قَالَ قَالَ رَسُولُ اللَّهِ صَلَّى اللَّهُ عَلَيْهِ وَسَلَّمَ مَنْ
اقْتَبَسَ عِلْمًا مِنْ النُّجُومِ اقْتَبَسَ شُعْبَةً مِنْ السِّحْرِ زَادَ مَا زَادَ.[51]

Anyone learning any part of astrology is learning a part of magic.

Abu Dawud also narrates that the Prophet prohibited ʿAli from mixing with astrologers.[52]

5: The Prophet forbade Muslims to deal with calculations in relation to Ramadan when he said that they were an unlettered people and that they neither wrote nor calculated. On the other hand, he told them to depend upon naked-eye sighting or complete thirty days. Some of them contend that the Prophet prohibited calculations because he was aware that Madinah's Jewish community was using them to confirm the Jewish months. In fact, the Jewish calendar was fixed by Rabbi Hillel II in 363 and the city's Jews had access to it. The Prophet intentionally prevented the Muslims from imitating them by ending the use of calculations to confirm the Islamic months.

حَدَّثَنَا آدَمُ حَدَّثَنَا شُعْبَةُ حَدَّثَنَا الأَسْوَدُ بْنُ قَيْسٍ حَدَّثَنَا سَعِيدُ بْنُ عَمْرٍو أَنَّهُ سَمِعَ ابْنَ عُمَرَ رَضِيَ اللَّهُ عَنْهُمَا عَنِ النَّبِيِّ صَلَّى اللَّهُ عَلَيْهِ وَسَلَّمَ أَنَّهُ قَالَ إِنَّا أُمَّةٌ أُمِّيَّةٌ لَا نَكْتُبُ وَلَا، نَحْسُبُ الشَّهْرُ هَكَذَا وَهَكَذَا يَعْنِي مَرَّةً تِسْعَةً وَعِشْرِينَ وَمَرَّةً ثَلَاثِينَ. ⁵³

We are an unlettered nation. We neither write nor calculate. The month is this way and this way. It means that sometimes it is twenty-nine days and sometimes thirty days.

6: Following astronomical calculations in matters of religion, such as the months of Ramadan and Shawwal, would nullify the spirit of the acts of worship like fasting. This contradicts the Prophet's clear commands, for he said, "Do not start fasting until you see the moon and do not stop fasting until you see the moon." He used both positive and negative verbal forms (fast by seeing and do not fast until you see it) to ensure that Muslims understood the significance of naked-eye sighting and did not follow Jewish practices in their faith and actions. Therefore, any Muslims who contravene his emphatic commands and start fasting based upon calculations must compensate for the days observed.

7: The Arabic word for the new moon is *hilāl*. The linguistic definition of this word requires that it deflect light and be shining, not dark. Shining, then, is connected with human sighting. Therefore, we cannot start the new month until we see the new moon.

CHAPTER 5

WEAKNESS OF THE ARGUMENT FOR *IKMĀL* (COMPLETING THIRTY DAYS)

So far we have analyzed the arguments based upon the Qur'an, Sunnah, and the linguistic implications of the word *al-hilāl.* Now, we will turn to the rest of the main arguments of those classical scholars who require naked-eye sighting.

Completing thirty days if the weather is cloudy is the generally agreed-upon position among the majority of classical scholars. Again, however, it is not the only categorical stance accepted by the Ummah, for various leading authorities (e.g., Ibn ʿUmar and Ahmad ibn Hanbal) are reported to have started the Ramadan fast after 29 Shaʿban if it was cloudy on the evening of 29 Shaʿban and thirty days had not been completed, as required by most of the prophetic narrations that include the phrase:

فإن غُبِّيَ عليكم فأكملوا عِدَّةَ شَعبانَ ثلاثين.

Complete counting thirty days of Shaʿban if it is obscured [from your eyes].

Ironically Ibn ʿUmar is the original narrator of most of these prophetic reports. In reality, the picture appears to be quite different when we discuss and analyze them in detail. In the following pages I will try to analyze some of them to prove that no such consensus exists, even on the rules governing the completion of the month. Several of the difficulties arising from these hadiths concerning completion can be appreciated only when we study these reports in depth and compare their final parts with one another.

حدّثنا آدمُ حدَّثَنا شُعبةُ حدَّثنا محمدُ بنُ زيادٍ قال: سمعتُ أبا هُريرةَ رضيَ اللّٰه عنه يقول: قال النبيُّ صلى الله عليه وسلم -أو قال: قال أبو القاسم صلى الله عليه وسلم- (صُوموا لِرُؤْيتِهِ وأفطِروا لرُؤيته، فإن غُبِّيَ عليكم فأكملوا عِدَّةَ شَعبانَ ثلاثين).[1]

The Prophet said, "Fast on sighting it [the new moon] and break the fast on sighting it. Complete thirty days of Shaʿban if it is cloudy."

وحدّثنا عُبَيْدُ اللّهِ بْنُ مُعَاذٍ. حَدَّثَنَا أَبِي. حَدَّثَنَا شُعْبَةُ عَنْ مُحَمَّدِ بْنِ زِيَادٍ قَالَ: سَمِعْتُ أَبَا هُرَيْرَةَ، رَضِيَ اللّهُ عَنْهُ يَقُولُ: قَالَ رَسُولُ اللّهِ (صُومُوا لِرُؤْيَتِهِ وَأَفْطِرُوا لِرُؤْيَتِهِ. فَإِنْ غُمِّيَ عَلَيْكُمُ الشَّهْرُ فَعُدُّوا ثَلاَثِينَ).[2]

The Prophet said, "Fast on sighting it [the new moon] and break the fast on sighting it. Count thirty days if the month is concealed from you [the sky being cloudy]."

Although the beginning of these reports is quite consistent in almost all of the hadiths, the part concerning completion varies. It seems that in this particular part the reporters are somehow explaining something, rather than just reporting the Prophet's exact words. Some of these reports are not as authentic as they seem to be.

In the two hadiths quoted above, it is important to note that both al-Bukhari and Muslim narrate from Abu Hurayrah through Muhammad ibn Ziyad. The first part of the hadith is the same in both narrations, but the ending is different in each. Al-Bukhari narrates:

Fa in ghubbiya ʿalaykum, fa akmilu ʿiddata Shaʿbāna thalāthīn.

فإن غُبِّيَ عليكم فأكملوا عِدَّةَ شعبانَ ثلاثين.

Whereas Muslim narrates:

Fa in ghummiya alaykum al-shahru, fa ʿuddu thalāthīn.

فَإِنْ غُمِّيَ عَلَيْكُمُ الشَّهْرُ فَعُدُّوا ثَلاَثِينَ.

The verb used by al-Bukhari is غُبِّيَ عليكم "*ghubbiya ʿalaykum*" (obscured from you); Muslim used a slightly different construction, "*ghummiya ʿalaykum al-shahru*" غُمِّيَ عَلَيْكُمُ الشَّهْرُ (if the month is obscured from you). In addition, al-Bukhari narrated "*fa akmilu ʿiddata Shāʿbāna thalāthīn*" فأكملوا عِدَّةَ شعبانَ ثلاثين (complete counting thirty days of Shaʿban); Muslim narrated "*fa ʿuddu thalāthīn*" (فَعُدُّوا ثَلاَثِينَ) ("then count thirty"), without using al-Bukhari's phrase ("then complete thirty days of Shaʿban"). Moreover, some of these narrations require the completion thirty days of Shaʿban only; the others require it for Ramadan as well.

Ahmad ibn Hanbal has reported a number of these narrations:

حدثنا عبد الله حدَّثني أبي ثنا إسماعيل أنا حاتم بن أبي صغيرة عن سماك بن حرب عن عكرمة قال: سمعت ابن عباس يقول: قال رسول الله صلى الله عليه وسلم: (صوموا لرؤيته وأفطروا لرؤيته، فإن حال بينكم وبينه سحاب فكمِّلوا العدة ثلاثين، ولا تستقبلوا الشهر إستقبالاً). قال حاتم: يعني عدة شعبان.[3]

The Prophet said, "Fast on sighting it [the new moon] and break the fast on sighting it. Complete thirty days [of Sha^c^ban] if the clouds come between you and the moon. And do not start the month ahead of time." Hatim said that it meant "counting thirty days of Sha^c^ban."

Here, Hatim seems to be giving his interpretation of the hadith as well.

حدثنا عبد الله حدَّثني أبي ثنا معاوية بن عمرو ثنا زائدة عن سماك بن حرب عن عكرمة عن ابن عباس قال: قال رسول الله صلى الله عليه وسلم: (صوموا لرؤيته وأفطروا لرؤيته، فإن حال دونه غيابة فأكملوا العدة، والشهر تسع وعشرون - يعني: إنه ناقص.[4]

Start fasting on sighting it and break the fast on sighting it. Complete the counting if the cloud covers it and the month is twenty-nine days, that is, it is incomplete.

Interestingly, both narrations are from Ibn ^c^Abbas. Ahmad ibn Hanbal used the same chain through Simak and ^c^Ikrimah; however, the narrations' endings are quite different. In the first one, the ending is:

(فإن حال بينكم وبينه سحاب فكمّلوا العدة ثلاثين ولا تستقبلوا الشهر إستقبالاً). قال حاتم: يعني عدة شعبان.[5]

If the clouds come between you and the moon, then complete counting thirty [days] and do not start the month ahead of time.

While in the second narration the ending is:

فإن حال دونه غيابة فأكملوا العدة، والشهر تسع وعشرون — يعني: إنه ناقص.

Complete the counting if the clouds cover it and the month is twenty-nine days, that is, it is incomplete.

In both of these narrations, the narrator is explaining the idea with the word يعني – "it means."

حدّثنا عبد الله، حدَّثني أبي، حدثنا يحي بن سعيد الأموي قال: ثنا الحجاج، عن عطاء، عن أبي هريرة قال: قال رسول الله صلى الله عليه وسلم: (صوموا لرؤيته وأفطروا لرؤيته، فإن غم عليكم الشهر، فأكملوا العدة ثلاثين).[6]

Fast on sighting it [the new moon], and break the fast on sighting it. Count thirty days if the month is concealed from you [the sky being cloudy].

حدّثنا عبد الله، حدَّثني أبي، حدثنا يحي بن سعيد، عن شعبة قال: ثنا محمد بن زياد، عن أبي هريرة، عن النبي صلى الله عليه وسلم قال: صوموا لرؤيته وأفطروا لرؤيته، فإن غم عليكم، فأكملوا العدة ثلاثين.[7]

Fast on sighting it [the new moon], and break the fast on sighting it. Complete thirty days [of Shaʿban] if it is cloudy.

حدَّثنا عبد الله، حدَّثني أبي، حدثنا حجاج قال: حدثنا شعبة، عن محمد بن زياد، قال: سمعت أبا هريرة قال: قال رسول صلى الله عليه وسلم: — أو قال: أبو القاسم عليه الصلاة والسلام: (صوموا لرؤيته، وأفطروا لرؤيته، فإن غم عليكم، فعدوا ثلاثين).[8]

Fast on sighting it [the new moon], and break the fast on sighting it. If it is cloudy, count thirty days.

Ahmad ibn Hanbal quotes these three narrations from Abu Hurayrah: two through Muhammad ibn Ziyad and one through ʿAta'. Interestingly enough, their endings also differ from those reported by al-Bukhari, who also uses narration from Abu Hurayrah through Muhammad ibn Ziyad. Al-Bukhari's version is:

Fa in ghubbiya ʿalaykum, fa akmilu ʿiddata Shaʿbāna thalāthīn.

فإن غُبِّيَ عليكم فأكملوا عِدَّةَ شَعبانَ ثلاثين.[9]

The verb is "*ghubbiya*" instead of "*ghumma*" and it is "*akmilu ʿiddata Shaʿbān thalāthīn*" instead of Ahmad ibn Hanbal's rendering of "*akmilu al-ʿiddata thalāthīn*" or "*fa ʿuddu thalāthīn.*"

Furthermore, in Ahmad ibn Hanbal, the three narrations from Abu Hurayrah show a little more variation. In the first two, the phrase فأكملوا العدة ثلاثين ("*fa akmilu*") is common to both. However, the first one says فإن غم عليكم الشهر ("*fa in ghumma ʿalaykum al-shahr*"); the second one says only فإن غم عليكم ("*fa in ghumma ʿalaykum*"). The difference between the second and the third report is that the second one says فأكملوا العدة ثلاثين ("*fa akmilu al-ʿiddata thalāthīn*"), whereas the third one says فعدوا ثلاثين ("*fa ʿuddu thalāthin*"). But both of them use فإن غم عليكم، ("*fa in ghumma ʿalaykum*").

حدّثنا عبد الله، حدَّثني أبي، حدثنا أبو عبد الرحمن عبد الله بن أحمد قال: حدَّثني أبي، قال: حدثنا محمد بن جعفر قال: حدثنا شعبة، عن محمد بن زياد، قال: سمعت أبا هريرة يقول: إن رسول الله صلى الله عليه وسلم قال: (لا تصوموا حتى تروا الهلال، ولا تفطروا حتى تروا الهلال، وقال: صوموا لرؤيته، وأفطروا لرؤيته، فإن غبي عليكم، فعدوا ثلاثين). شعبة، وأكثر علمي أنه قال: لا تصوموا حتى تروا الهلال، ولا تفطروا حتى تروا الهلال.[10]

The Prophet said, "Do not fast until you see the crescent [moon] and do not break the fast until you see the crescent [moon]." And he said, "Fast on seeing it and break the fast on seeing it. If there is confusion, then count thirty days."

Shu῾bah said: "To the best of my knowledge, he said, 'Do not fast until you see the [new] moon and do not break the fast until you see the [new] moon.'"

Shu῾bah's comments at the end are not clear when he says "he said." To whom does "he" refer – the Prophet, Muhammad ibn Ziyad, or Abu Hurayrah?

Shu῾bah's statement "to the best of my knowledge" and his emphsis upon the first portion of the hadith "Do not fast until you see the [new] moon and do not break the fast until you see the [new] moon," also clearly indicates that perhaps he is aware of such differences in the endings of various narrations. There seems to be agreement on their beginning, both the positive and negative renderings; however, the endings are somewhat problematic in that they vary from one narration to another. In addition, most of them do not agree with al-Bukhari's rendering but, to a greater extent, with that of Muslim.

The following narrations present additional variations in the text of these reports.

قال عبد الله: وجدت هذين الحديثين في كتاب أبي بخط يده قال: حدثنا محمد بن عبد الله الأنصاري، حدثنا محمد بن عمرو، عن أبي سلمة، عن أبي هريرة قال: قال: رسول الله صلى الله عليه وسلم: (لا تقدموا الشهر ــ يعني رمضان ــ بيوم ولا يومين إلا أن يوافق ذلك صوماً كان يصومه أحدكم، صوموا لرؤيته، وأفطروا لرؤيته، فإن غم عليكم فعدوا ثلاثين، ثم أفطروا).[11]

The Prophet said, "Do not start the month – meaning Ramadan (*ya῾nī Ramaḍān*) – a day or two ahead of time [unless it coincides with one's routinely observed days of fasting]. Observe the fast on sighting it [the new moon] and break the fast on sighting it. Count thirty days if it is cloudy, then break the fast."

The use of يعني رمضان ("*ya῾nī Ramaḍān*") in this narration indicates that the narrator is explaining something or giving his personal understanding of the issue and that it is not just confined to transmitting the original text. Moreover, this narration adds that Ramadan should also be counted as thirty days, which is not mentioned in the narrations discussed earlier.

حدّثنا عبد الله حدّثني أبي ثنا إسحاق بن عيسى أنا محمد بن جابر عن قيس بن طلق عن أبيه قال: قال رسول الله صلى الله عليه وسلم: (إن الله عزَّ وجلَّ جعل هذه الأهلة مواقيت للناس، صوموا لرؤيته وافطروا لرؤيته، فإن غم عليكم فأتموا العدة).[12]

On the authority of Qays ibn Talq, who reported that his father said, "The Prophet said, 'Truly, Allah has made these new moons as signs to mark fixed periods of time for [the affairs of] people. Fast sighting it and break the fast on sighting it, and complete the number [of thirty days] if it is cloudy.'"

حدّثنا عبد الله، حدَّثني أبي، حدثنا يحيى، عن شعبة، قال: حدثنا محمد بن زياد، عن أبي هريرة، عن النبي صلى الله عليه وسلم قال: (صوموا لرؤيته وأفطروا لرؤيته، وإن غم عليكم، فأكملوا العدة ثلاثين).[13]

Abu Hurayrah reported that the Prophet said, "Fast on sighting it [the new moon] and break the fast on sighting it. Complete thirty days [of Shaʿban] if it is cloudy."

حدّثنا عبدالله حدَّثني أبي حدثنا يحيى بن زكريا قال: أنبانا حجاج عن حسين بن الحارث الجدلي قال: خطب عبدالرحمن بن زيد بن الخطاب في اليوم الذي يشك فيه، فقال: ألا أني قد جالست أصحاب رسول الله صلى الله عليه وسلم وسألتهم، ألا وأنهم حدَّثوني أن رسول الله صلى الله عليه وسلم قال: صوموا لرؤيته وافطروا لرؤيته، وإن تشكوا لها فإن غم عليكم فأتموا الثلاثين، وإن شهد شاهدان مسلمان فصوموا وأفطروا.[14]

ʿAbd al-Rahman ibn Zayd ibn al-Khattab gave a *khutbah* on the day of doubt and said, "I sat with the Companions of the Prophet and asked them, and they told me that the Prophet said, 'Fast on sighting it, and break the fast on sighting it. If you are in doubt, then complete [*atimmu*] the thirty [days of Shaʿban] if it is cloudy. But if two Muslim witnesses testify [that they saw the new moon], then fast or break your fast accordingly.'"

Several additional points are discussed in this narration. It addresses the issue of fasting on the day of doubt and uses وإن تشكوا لها. Instead of *akmilu* or *ʿuddu*, it uses *atimmu*. Finally, it clearly addresses the *fiqhī* discussion of whether the month is confirmed by the witness of one or two Muslims and sides with those jurists who accept the sighting of two witnesses rather than one, as is the opinion of Imam Malik. ʿAbd al-Rahman ibn Zayd ibn al-Khattab does not mention the names of the Companions from whom he is narrating. He only reports that he heard it from some of them who had narrated it to him.

حدّثنا عبد الله، حدَّثني أبي، حدثنا حماد، حدثنا حماد بن سلمة، أنبانا محمد بن زياد قال: سمعت أبا هريرة، يقول: سمعت أبا القاسم صلى الله عليه وسلم يقول: صوموا الهلال لرؤيته، وأفطروا لرؤيته، فإن غم عليكم فعدوا ثلاثين.[15]

Muhammad ibn Ziyad narrated that he heard Abu Hurayrah saying, "I heard Abu al-Qasim [that is, the Prophet – peace and blessings be upon him] saying, 'Fast on sighting the new moon [*al-hilāl*], and break your fast on sighting it, and count thirty [days] if it is cloudy [*ghumma ʿalaykum*].'"

It is significant that although Muhammad ibn Ziyad narrates it from Abu Hurayrah, the same chain previously discussed in al-Bukhari's narration, the text is not exactly the same. It adds الهلال before لرؤيته, uses غم عليكم instead of غُبِّيّ عليكم, and finally, uses فعُدُّوا ثلاثين instead of al-Bukhari's version: فأكملوا عِدَّةَ شَعبانَ ثلاثين. It also differs from Muslim's narration in that it puts الهلال before لرؤيته, a word that does not occur in Muslim's narration. The second is that it uses غم عليكم instead of Muslim's غُمِّيَ عَلَيْكُمُ الشَّهْرُ.

حدّثنا عبد الله حدَّثني أبي حدثنا سليمان بن داود الطيالسي ـــ أبو داود ـــ أنبانا عمران عن قتادة عن الحسن عن أبي بكرة أن النبي صلى الله عليه وسلم قال: ـــ يعني ـــ "صوموا الهلال لرؤيته، وأفطروا لرؤيته، فإن غمّ عليكم، فأكملوا العدة ثلاثين، والشهر هكذا وهكذا، وهكذا..."[165]

Al-Hasan reported from Abu Bakr that the Prophet explained what he meant (*yaʿnī*), "Fast on sighting the new moon [*al-hilāl*] and break your fast on sighting it, and complete the number [of thirty days] if it is cloudy. The month is thus and thus [that is, it is either 29 or 30 days]," and he [the Prophet] indicated [this] with his fingers.

The use of يعني ("which means") in this case indicates that the narration is not the exact transmission of the original statement, but a somewhat modified version. It also adds "*wa al-shahru hākathā, wa hākathā, wa hākathā, wa ʿaqad*" (والشهر هكذا وهكذا، وهكذا، وعقد).

حدثنا أبو كُرَيْبٍ حدثنا عَبْدَةُ بنُ سُلَيْمَانَ عن محمدِ بنِ عَمْرٍو عن أبي سَلَمَةَ عن أبي هريرةَ، قال: قال النبيُّ "لا تَقَدمُوا الشَّهْرَ بِيَوْمٍ ولا بِيَوْمَيْنِ إلا أنْ يُوَافِقَ ذلِكَ صَوْماً كانَ يَصُومُهُ أَحَدُكُم. صُومُوا لِرُؤْيتِهِ وأَفْطِرُوا لِرُؤْيتِهِ فإن غُمَّ عَلَيْكُم فعُدُّوا ثلاثينَ ثُمَّ أفْطِرُوا." قال: وفي البابِ عن بعضِ أصحابِ النبيِّ (أخبرنا مَنْصُورُ بنُ الْمُعْتَمِرِ عن رِبْعِيِّ بنِ حِرَاشٍ عن بعضِ أصحابِ النبيِّ عَنِ النَّبيِّ بِنَحْوِ هذا. قال أبو عيسى: حديثُ أبي هريرةَ حديثٌ حسنٌ صحيحٌ. والعملُ على هذا عِنْدَ أهلِ العلمِ: كَرِهُوا أن يَتَعَجَّلَ الرَّجُلُ بِصِيَامٍ قَبْلَ دُخُولِ شهْرِ رَمَضَانَ لِمَعْنَى رَمَضَانَ وإنْ كانَ رَجُلٌ يَصُومُ صَوْماً فَوَافَقَ صِيَامُهُ ذلِكَ فلا بأسَ به عندَهُم.[17]

Abu Hurayrah reported that the Prophet said, "None of you should fast a day or two before Ramadan unless it coincides with a day that a person customarily fasts. Fast when you see the new moon and break it when you see it. If cloud obscures it, then complete the thirty days [of Ramadan], then break your fast."

This report emphasizes completing the thirty days of Ramadan, rather than of Sha^cban, as is the case with the majority of narrations already mentioned. It also deals with the question of whether to fast on the day of doubt.

حدثنا قُتَيْبَةُ، حدثنا أبو الأحْوَصِ عن سِمَاكِ بن حَرْبٍ عن عِكْرِمَةَ عن ابنِ عباسٍ، قال: قال رسولُ الله (لا تَصُومُوا قَبْلَ رَمَضَانَ، صُومُوا لِرُؤْيَتِهِ وأفْطِرُوا لِرُؤْيَتِهِ، فإنْ حَالَتْ دُونَهُ غَيَابَةٌ فأكْمِلُوا ثلاثين يَوْماً). وفي البابِ عن أبي هريرةَ وأبي بَكْرَةَ وابنِ عُمَرَ. قال أبو عيسى: حديثُ ابنِ عبَّاسٍ حديثٌ حسنٌ صحيحٌ. وقد رُوِيَ عنْهُ مِنْ غَيْرِ وَجْه.[18]

Qutaybah narrated from Abu al-Ahwas, from Simak ibn Harb, who reported that ʿAbd Allah ibn ʿAbbas said: "The Prophet said, 'Do not fast immediately before Ramadan. Start the fast with the sighting of the new moon and break your fast on sighting it. If the sky is overcast, then complete thirty days.'"

حدَّثنا عبدُاللَّهِ بنُ سعيدٍ ثَنا إسماعيلُ بنُ عليةَ، ثَنا حاتمُ بنُ أبي صغيرةَ، عن سماكِ بنِ حربٍ، قالَ: أصبحتُ في يومٍ قد أشكلَ عليَّ من شعبانَ أو من شهرِ رمضانَ، فأصبحتُ صائماً فأتيتُ عكرمةَ فإذا هوَ يأكُلُ خبزاً وبقلاً، فقالَ: هلمَّ إلى الغداءِ فقلتُ: إني صائمٌ فقالَ: أُقْسِمُ باللَّهِ لَتُفْطِرَنَّ، فلما رأيتُه حلفَ ولا يستثني تقدَّمْتُ فعذّرت وإنما تسحرتُ قبيلَ ذَلِكَ ثُمَّ قلتُ هاتِ الآنَ ما عندَكَ فقالَ: حدَّثَنا ابنُ عباسٍ قالَ: قالَ رسولُ اللَّهِ صلى الله عليه وسلم: (صوموا لرؤيتِهِ وأفطِروا لرؤيتِهِ، فإنْ حالَ بينَكم وبينَهُ سحابٌ فكمّلوا العدةَ ثلاثينَ، ولا تستقبلوا الشهرَ استقبالاً).[19]

Simak ibn Harb reported: "I woke up one day feeling confused about whether the day was [the last] day of Shaʿban or [the first of] Ramadan, so I started the day fasting. Then I went to ʿIkrimah and saw him eating bread and herbs. He said, 'Come and have lunch with me,' so I told him I was fasting. Thereupon he said, 'I swear by Allah that you must break your fast.' When I saw him swearing without making any exception, I went forward and asked about his proof. He said, 'Ibn ʿAbbas reported that the Prophet said, "Fast on sighting it [the new moon], and break your fast on sighting it. Complete thirty days if it is cloudy, and do not immediately fast before the month [of Ramadan] starts".'"

حدثنا مُحمَّدُ بنُ الصَّباحِ الْبَزَّازُ أخبرنا جَرِيرُ بنُ عَبْدِ الْحَمِيدِ الضَّبِّيُّ عَنْ مَنْصُورِ بن الْمُعْتَمِرِ عن رِبْعِيِّ بن حِراشٍ عن حُذَيْفَةَ، قال: قال رَسُولُ الله صلى الله عليه وسلم: (لَا تُقَدِّمُوا الشَّهْرَ حتى تَرَوْا الهِلَالَ أوْ تُكْمِلُوا الْعِدَّةَ ثُمَّ صُومُوا حتى تَرَوْا الهِلَالَ أوْ تُكْمِلُوا الْعِدَّةَ). قالَ أَبو دَاوُدَ: رَوَاهُ سُفْيَانُ وَغَيْرُهُ عن مَنْصُورٍ عن رَبْعِيٍّ عن رَجُلٍ من أصحابِ النَّبِيِّ صلى الله عليه وسلم لَمْ يُسَمِّ حُذَيْفَةَ.[20]

On the authority of Hudhayfah, who reported that the Prophet said, "Do not fast a day or two before Ramadan until you see the new moon or complete thirty days [of Shaʿban]. Then fast until you see the [new] moon [of Shawwal] or complete thirty days." Abu Dawud adds: "This hadith has been narrated by Sufyan and others on the authority of Mansur, who reported it from Rabiʿi from a Companion of the Prophet and he did not name Hudhayfah."

It is clear from the above narrations that various other alterations were made to the narration of the same Companion from the same chain. All of these narrations agree upon the positive or negative formula of صوموا لرؤيتِهِ وأفطِروا لرؤيتِهِ; however, they differ markedly in the other aspects of the narration. Some of the narrators are not even sure of the Companions' name, or at least they do not mention some of their names, as in the report from Hudhayfah, for instance, in the above narration by al-Darimi.

أخبرنا الحسينُ بنُ إدريس الأنصاريُّ، قال: حدَّثنا عثمانُ بنُ أبي شيبةَ، قال: حدَّثنا جريرٌ، عن منصورٍ، عَنْ ربعيِّ بنِ حِراشٍ عن حذيفةَ، قال: قَالَ رَسُولُ اللَّه (لا تَقَدَّمُوا الشَّهْرَ حتَّى تَرَوُا الهِلالَ، أو تُكمِلُوا العِدَّةَ، ثُمَّ صُوموا حتَّى تَرَوُا الهِلال أو تُكْمِلُوا العِدَّةَ).[21]

Rabiʿi ibn Jarrash reported on the authority of Hudhayfah, who narrated that the Prophet said, "Do not fast immediately before the month [of Ramadan] until you see the new moon or complete the number [of thirty days of Shaʿban]. Then fast until you see the new moon or complete the thirty days."

وأخبرنا أبو علي الرُّوْذَبارِيُّ أنبأ محمد بن بكر ثنا أبو داودَ حدثنا الحسن بن علي ثنا حسين عن زائدة عن سِمَاكٍ عن عِكْرِمَةَ عن ابنِ عباسٍ قالَ: قالَ رسولُ اللهِ صَلَّى اللهُ عَلَيْهِ وَسَلَّمَ: (لَا تَقَدَّمُوا الشهرَ بصيامِ يومٍ ولا يومينِ، إلَّا أَنْ يكونَ شيئاً يصومُهُ أحدُكُمْ، ولا تصومُوا حتَّى تَرَوْهُ، ثم صومُوا حتَّى تَروُهُ، فإنْ حالَ دونَهُ غمامَةٌ، فأتِمُّوه العِدَّةَ ثلاثينَ ثم أَفْطِرُوا ثم أَفْطِرُوا، الشهرُ تِسع وعشرونَ). قال أبو داودَ: ورواه حاتِمُ بنُ أبِي صَغِيْرَةَ وشعبةُ والحسنُ بن صالح عن سِمَاكٍ بمعناهُ لم يقولوا: (ثم افطروا). (قال الشيخ) ورواه أبو عَوَانَةَ عن سِمَاكٍ مختصراً فَجَعَلَ إكْمالَ العِدَّةِ لشعبانَ.[22]

Ibn ʿAbbas reported that the Prophet said, "Do not fast a day or two before the month [of Ramadan] unless it is a day on which a person customarily fasts. And do not fast until you see the new moon. Then fast until you see it. But if a cloud obscures it, then complete the number of thirty days, then break your fast, then break your fast – the month has twenty-nine days."

SIGNIFICANCE OF THE VARIATIONS IN TRANSMISSION

I have included most of the reports narrated in the books of hadith regarding this question to show the variety as well as the level of variance among them. It is clear from the hadiths quoted above that there is almost unanimous agreement on the beginning: "Fast on sighting it and break your fast on sighting it, (صوموا لرؤيتِهِ وأَفطِروا لرؤيتِهِ)، . But their endings vary considerably, even those from the same original narrator or the same chain. Therefore, as Ahmad Shafat points out, it is possible that the "amount of variation in language shows that the transmitters of the hadith are describing an idea freely in their own words rather than attempting to transmit the hadith with faithfulness to the original words."[23]

Completing thirty days of Shaʿban or of Ramadan in the case of obscurity is the adopted opinion of the majority of jurists *(al-jumhūr)*. In their view, there are only two methods of confirming the Islamic month: naked-eye sighting or completion. It is useful to note here that there is almost unanimous agreement on the part of the hadiths referring to the sighting of the moon (in positive as well as in frequently negative terms). The only part that presents a wide range of variations is that referring to completion, as seen above. However, these are exactly the same parts upon which most scholars base their arguments for explaining away the more authentic rendering from Ibn ʿUmar – "*faqduru lahu*," as will be discussed below. They contend that the meaning of فَإِنْ غُمَّ عَلَيْكُمْ فَاقْدُرُوا لَهُ "*fa in ghumma ʿalaykum faqduru lahu*" in Ibn ʿUmar's narration is "complete thirty days and do not follow counting or calculations," as the literal meaning of the narration apparently demands. Al-Nawawi contends:

وَاحْتَجَّ الْجُمْهُورُ بِالرِّوَايَاتِ الَّتِي ذَكَرْنَاهَا وَكُلُّهَا صَحِيحَةٌ صَرِيحَةٌ: فَأَكْمِلُوا الْعِدَّةَ ثَلاثِينَ وَاقْدُرُوا لَهُ ثَلاثِينَ، وَهِيَ مُفَسِّرَةٌ لِرِوَايَةِ فَاقْدُرُوا لَهُ الْمُطْلَقَةِ.[24]

"The majority (*al-jumhūr*) has derived from the clear and authentic prophetic narrations quoted above that completing thirty days and counting thirty days [are the same]. The explaining phrase (مُفَسِّرَةٌ) "complete thirty days" explains away the general phrase (المُطلقة) "then count or estimate it."

Additionally, there is no consensus among the jurists even on this interpretation of فَاقْدُرُوا لَهُ "*faqduru lahu*," for Ahmad ibn Hanbal argues that this particular prophetic phrase means "shorten the month."

(فَإِنْ غُمَّ عَلَيْكُمْ فَاقْدُرُوا لَهُ) فَقَالَ أَحْمَدُ بْنُ حَنْبَلٍ وَطَائِفَةٌ قَلِيلَةٌ: مَعْنَاهُ ضَيِّقُوا لَهُ وَقَدِّرُوهُ تَحْتَ السَّحَابِ.[25]

"Calculate (estimate) it if it is obscured from you." Ahmad ibn Hanbal along with a minority group said that it meant restricting it and considering (the moon) under the clouds.

Al-Nawawi himself reports that Ahmad ibn Hanbal and a few others say that the meaning is not to complete thirty days, but "restrict it or shorten the month by considering the moon behind the clouds." That is why the later contends that fasting should be observed the next day, the day after 29 Shaʿban, if, owing to obscurity, the new moon is not sighted on that particular evening. Abu Dawud reports that this was the opinion and practice of Ibn ʿUmar.

حدثنا سُلَيْمانُ بنُ دَاوُدَ الْعَتَكِيُّ أخبرنا حَمَّادٌ أخبرنا أَيُّوبُ عن نَافِعٍ عن ابن عُمَرَ، قال قال رَسُولُ الله صلى الله عليه وسلم: الشَّهْرُ تِسْعٌ وَعِشْرُونَ فَلَا تَصُومُوا حَتَّى تَرَوْهُ وَلَا تُفْطِرُوا حَتَّى تَرَوْهُ. فَإِنْ غُمَّ عَلَيْكُمْ فَاقْدُرُوا لَهُ ثَلَاثِينَ. قالَ: فَكَانَ ابنُ عُمَرَ إذَا كان شَعْبَانُ تِسْعاً وَعِشْرِينَ نُظِرَ لَهُ فإن رُئِيَ فَذَاكَ وَإن لَم يُرَ وَلَمْ يَحُلْ دُون مَنْظَرِهِ سَحَابٌ وَلَا قَتَرَةٌ أَصْبَحَ مُفْطِراً، فَإِنْ حَالَ دُونَ مَنْظَرِهِ سَحَابٌ أَوْ قَتَرَةٌ أَصْبَحَ صَائِماً. قال وَكَانَ ابنُ عُمَرَ يُفْطِرُ مَعَ النَّاسِ وَلَا يَأْخُذُ بِهذَا الحِسَابِ.[26]

Ibn ʿUmar narrates that the Prophet said, "The month is twenty-nine [days], therefore do not start fasting until you see it and do not break the fast until you see it. Count thirty days if it is cloudy." He [Nafiʿ, the narrator from Ibn ʿUmar] said that Ibn ʿUmar used to ask people to look for the [new] moon for him on 29 Shaʿban. If the moon was seen, then well and good. He would not fast if it was not seen and it was not cloudy or rainy. If [on 29 Shaʿban] it was cloudy or rainy weather, he would start fasting [the next day]. He [Nafiʿ] also said that Ibn ʿUmar used to break the fast with the people and not depend upon these calculations [his counting].

Al-Bayhaqi also reports the same:

أخبرنا أبو عبد الله الحافظ ثنا محمد بن يعقوبَ هو الشَّيْبَانِيُّ ثنا محمد بن شَاذَانَ الأَصَمُّ ثنا عليُّ بنُ حُجْرٍ ثنا إسمعيلُ عن أيوبَ (ح وأخبرنا) أبو الحسن علي بن محمد المقرئُ أنبأ الحسن بن محمد بن إسحاقَ ثنا يوسفَ بنُ يعقوبَ ثنا سليمانُ بن حرب ثنا حماد بن زيد عن أيوبَ عن نافعٍ عن ابنِ عُمَرَ قالَ: قالَ رسولَ الله صَلَّى الله عَلَيْهِ وَسَلَّمَ: (إنَّمَا الشَّهْرُ تِسْعٌ وعشرونَ فلا تَصُوْمِوا حتَّى تَرَوْهُ ولا تُفْطرُوا حتَّى تَرَوْهُ، فإنْ غُمَّ عَلَيْكُمْ فاقْدِرُوا لَهُ).[27]

Ibn ʿUmar reported that the Prophet said, "Truly, the month consists of twenty-nine days. So do not fast until you see it, do not break your fast until you see it, and estimate it if it is cloudy."

زاد حماد في روايتهِ عن أيوبَ: قالَ نافعٌ: كانَ ابنُ عُمَرَ إذَا مَضَى من شعبانَ تِسْعٌ وعشرونَ نُظِرَ لَهُ فإنْ رُئِيَ فداكَ، وإنْ لم يُرَ ولم يَحُلْ دُوْنَ مَنْظَرِهِ سحابٌ ولا قَتَرَةٌ أصْبَحَ مُفْطِراً، وإن حالَ دُوْنَ مَنْظَرِهِ سحابٌ أو قَتَرَةٌ أصْبَحَ صائماً، وكانَ يُفْطِرُ معَ الناسِ ولا يَأْخُذُ بِهَذَا الحسابِ.[28]

Hammad added in his report from Ayyub that Nafᶜi said Ibn ᶜUmar used to ask people to look for the [new] moon for him on 29 Shaᶜban. If it was seen, then well and good. He would not fast if it was not seen and it was not cloudy or rainy. If [on 29 Shaᶜban] it was cloudy or rainy weather, he would start fasting [the next day]. He [Nafᶜi] also said that Ibn ᶜUmar used to break the fast with the people and not depend upon these calculations [his counting].

This narration is important from a number of perspectives. First, it is quoted to prove that Ibn ᶜUmar did not rely on calculations. It seems that some of the jurists have explained the sentence "*wa lā ya'khuthu bihāthā al-ḥisāb*" (he did not depend upon these calculations) ولا يَأْخُذُ بِهَذَا الحسابِ to mean that Ibn ᶜUmar did not follow calculations. This interpretation is incorrect, for al-Azimabadi clearly shows that he used to break the fast with the rest of the Muslims and not worry about his calculations of the day upon which he had started his fast. If Ramadan was twenty-nine days, then his starting day would be the thirtieth day for him. If Ramadan turned out to be thirty days, then he would consider his first day as supplementary fasting for Shaᶜban. This is the correct interpretation of the phrase.

Second, this is Ibn ᶜUmar's only narration on the subject of our discussion that includes فَاقْدُرُوا لَهُ ثَلاَثِيْنَ (then count or estimate it as thirty days). Later, we will see that this is the only narration from him that requires counting thirty days in the case of cloudy weather. All of the other reports confine themselves to فَاقْدُرُوا لَهُ "*faqduru lahu*" (restrict it, calculate it) and are explained by the *jumhūr* in light of this oddly attached report. This narration even contradicts itself, for Ibn ᶜUmar's action is posted against his own narration that states "complete thirty days if it is cloudy." He started fasting after completing only twenty-nine days of Shaᶜban when there was an obscurity on the horizon.

Ibn Qudamah argues that Ibn ᶜUmar has explained the true meanings of the hadith by his own action and that must be taken as incumbent, for he is the original reporter of the hadith requiring Muslims to estimate in the case of cloudy weather and complete thirty days.

وَمَعْنَى اُقْدُرُوا لَهُ: أَيْ ضَيِّقُوا لَهُ الْعَدَدَ مِنْ قوله تعالى: (وَمَنْ قُدِرَ عَلَيْهِ رِزْقُهُ). أَيْ ضُيِّقَ عَلَيْهِ. وَقَوْلِهِ: (يَبْسُطُ الرِّزْقَ لِمَنْ يَشَاءُ وَيَقْدِرُ). وَالتَّضْيِيقُ لَهُ أَنْ يُجْعَلَ شَعْبَانُ تِسْعَةً وَعِشْرِينَ يَوْمًا. وَقَدْ فَسَّرَهُ ابْنُ عُمَرَ بِفِعْلِهِ ، وَهُوَ رَاوِيه ، وَأَعْلَمُ بِمَعْنَاهُ ، فَيَجِبُ الرُّجُوعُ إِلَى تَفْسِيرِه.[29]

The meaning of "calculate it" is to restrict the counting of it. Allah has said in the Qur'an: "the one whose sustenance was restricted," meaning reduced or shortened.

[There is] also His statement that "it is Allah who increases the sustenance for those whom He likes and reduces [it] for those whom He dislikes." Shortening or reducing in the case of cloudy weather will mean to make Sha'ban twenty-nine days. Ibn 'Umar has explained the meaning of the hadith by his practice. He is the original narrator of this report and is better equipped to understand its true meaning. Therefore, it is obligatory to return to his explanation.

قَالَ أَبُو مُحَمَّدٍ: هَذَا ابْنُ عُمَرَ هُوَ رَوَى أَنْ لا يُصَامُ حَتَّى يَرَى الْهِلالَ ثُمَّ كَانَ يَفْعَلُ مَا ذَكَرْنَا.[30]

Abu Muhammad ibn Hazm said, "This is Ibn 'Umar, who himself narrated that fasting was not permitted until the new moon was sighted, then he himself does what we have just mentioned."

Was Ibn 'Umar contravening the Prophet's command, which he himself narrated, by fasting on the day after 29 Sha'ban in the case of obscurities? It is reported that many Companions and Successors fasted in the case of obscurities so as not to miss a day of Ramadan. Ibn Qudamah reports:

مَسْأَلَةٌ: قَالَ: (وَإِنْ حَالَ دُونَ مَنْظَرِهِ غَيْمٌ، أَوْ قَتَرٌ وَجَبَ صِيَامُهُ، وَقَدْ أَجْزَأَ إِذَا كَانَ مِنْ شَهْرِ رَمَضَانَ) اخْتَلَفَتِ الرِّوَايَةُ عَنْ أَحْمَدَ رحمه الله فِي هَذِهِ الْمَسْأَلَةِ، فَرُوِيَ عَنْهُ مِثْلُ مَا نَقَلَ الْخِرَقِيُّ، اخْتَارَهَا أَكْثَرُ شُيُوخِ أَصْحَابِنَا، وَهُوَ مَذْهَبُ عُمَرَ، وَابْنِهِ، وَعَمْرِو بْنِ الْعَاصِ، وَأَبِي هُرَيْرَةَ، وَأَنَسٍ، وَمُعَاوِيَةَ، وَعَائِشَةَ، وَأَسْمَاءَ بِنْتَيْ أَبِي بَكْرٍ، وَبِهِ قَالَ بَكْرُ بْنُ عَبْدِ اللَّهِ، وَأَبُو عُثْمَانَ النَّهْدِيُّ، وَابْنُ أَبِي مَرْيَمَ، وَمُطَرِّفٌ، وَمَيْمُونُ بْنُ مِهْرَانَ، وَطَاوُسٌ، وَمُجَاهِدٌ.[31]

If any obscurity such as a cloud or rain comes between the new moon and its sighting, then fasting is obligatory. This day will be accepted if it was the first day of Ramadan. There are contradictory reports from Ahmad about this issue. The report that is transmitted by al-Kharqi from Imam Ahmad is accepted by the majority of the teachers of our scholarship. That is also the preferred opinion of Caliph 'Umar and his son ['Abd Allah], 'Umar ibn al-'As, Abu Hurayrah, Anas, Mu'awiyah, 'A'ishah, Asma' (the two daughters of Abu Bakr). The same opinion was held by Bakr ibn 'Abd Allah, Abu 'Uthman al-Nahdi, Ibn abi Maryam, Mutarraf, Maymun ibn Mehran, Tawus, and Mujahid.

Al-Nawawi reports that Ahmad ibn Hanbal requires the starting of Ramadan if the new moon has not been sighted owing to obscurities on 29 Sha'ban. That has been the opinion of eight known Companions and seven of the Successors, including 'Umar and his son Ibn 'Umar.[32] Al-Azimabadi states:

وروي معناه عن أبي هريرة وابن عباس رضي الله عنهما وعائشة وأسماء ابنتا أبي بكر تصومان ذلك اليوم، وقالت عائشة رضي الله عنها: لأن أصوم يوماً من شعبان أحب إليَّ من أن أفطر يوماً من رمضان. وكان مذهب عبد الله بن عمر بن الخطاب رضي الله عنهما صوم يوم الشك إذا كان في السماء سحاب أو قترة، فإن كان صحو ولم ير الناس الهلال أفطر مع الناس، وإليه ذهب أحمد بن حنبل.[33]

Its meanings have been narrated from Abu Hurayrah and Ibn ʿAbbas. ʿA'ishah and Asma', Abu Bakr's two daughters, used to fast this day. ʿA'ishah used to say, "It is better for me to fast a day of Shaʿban than miss a day of Ramadan." ʿAbd Allah ibn ʿUmar used to fast on the day of doubt if there were obscurities on the horizon, such as clouds or rain. He would not fast if the horizon was free of obscurities and the people were unable to sight the moon. Ahmad's opinion is the same.

Some scholars have argued that Ibn ʿUmar and others used to fast on the day of doubt with the intention of engaging in supplementary fasting, and thereforoe not considering it as a day of Ramadan. This interpretation is also incorrect, as Ahmad ibn Hanbal clearly reports:

يجب صومه على أنه من رمضان.[34]

It is obligatory to fast that day as the first day of Ramadan.

The "day of doubt" is defined by al-ʿAyni as:

قال العلامة العيني: ويوم الشك هو اليوم الذي يتحدث الناس فيه برؤية الهلال ولم يثبت رؤيته أو شهد واحد فردت شهادته أو شاهدان فاسقان فردت شهادتهما.[35]

The "day of doubt" is the day when people talk about sighting the [new] moon, yet its sighting is not confirmed. For instance, it was reported by only one witness and his report was denied, or two untrustworthy individuals reported [it] and their statement was rejected.

Third, the report establishes that the original narrator, Ibn ʿUmar, did not accept the explanatory note as "complete thirty days," for he contravened it and fasted after 29 Shaʿban in the case of obscurities. How can anyone assert that there is a consensus among the *jumhūr* that "complete thirty days" is the true meaning of "count it?" In fact, there was no consensus even among the Companions or their Successors that "complete thirty days" explains "count or estimate it." Had it been an accepted norm, as al-Nawawi and many others contend, then Ibn ʿUmar, ʿA'ishah, Asma', Imam Ahmad, and many others would have not violated it by fasting after 29 Shaʿban in the case of cloudy weather.

Fourth, the practice of Ibn ʿUmar and Ahmad of fasting on the cloudy twenty-ninth day without having physically sighted the new moon refutes the argument of the majority (*jumhūr*) that either naked-eye sighting or completing thirty days is the only prescribed method for confirming all of the Islamic months, including Ramadan. Ibn ʿUmar or Ahmad ibn Hanbal started Ramadan by counting the twenty-nine days of Shaʿban. This method of confirming the month, when it is cloudy, on 29 Shaʿban is not naked-eye sighting or completion, but "counting the days."

Fifth, al-Bukhari, Muslim, and others record a hadith that the Prophet himself started or ended the month without resorting to naked-eye sighting or completing thirty days.

حَدَّثَنَا أَبُو عَاصِمٍ عَنْ ابْنِ جُرَيْجٍ عَنْ يَحْيَى بْنِ عَبْدِ اللَّهِ بْنِ صَيْفِيٍّ عَنْ عِكْرِمَةَ بْنِ عَبْدِ الرَّحْمَنِ عَنْ أُمِّ سَلَمَةَ رَضِيَ اللَّهُ عَنْهَا أَنَّ النَّبِيَّ صَلَّى اللَّهُ عَلَيْهِ وَسَلَّمَ آلَى مِنْ نِسَائِهِ شَهْرًا فَلَمَّا مَضَى تِسْعَةٌ وَعِشْرُونَ يَوْمًا غَدَا أَوْ رَاحَ فَقِيلَ لَهُ إِنَّكَ حَلَفْتَ أَنْ لَا تَدْخُلَ شَهْرًا فَقَالَ إِنَّ الشَّهْرَ يَكُونُ تِسْعَةً وَعِشْرِينَ يَوْمًا.[36]

Umm Salamah narrates that once the Prophet took an oath not to see his wives for a month. When the twenty-nine days had passed, he came to them. He was told: "You took an oath not to enter the home for a month." He said, "The month consists of twenty-nine days."

The Prophet just counted the days and completed his month or started the new month without seeing the new moon. He did not say that he had seen it, and none of his wives asked him if he had seen it. The hadith does not say that it was cloudy that evening, and the Prophet clearly did not complete the thirty days. Imam al-Bukhari narrates the same report from a different narrator, Anas ibn Malik,[37] and al-Tirmidhi has authenticated it.[38] It seems that the Prophet determined the month just by counting twenty-nine days.

By now, it should be sufficiently clear that the assertion that the entire Ummah or all jurists have reached a consensus that Islamic lunar months cannot be determined without naked-eye sighting or completion is both inauthentic and untrue, for many instances of practical exceptions to this rule are substantiated by the points already discussed. Moreover, both interpretations of "completing the thirty days of Shaʿban and Ramadan in the case of cloudy weather" (the *jumhūr's* opinion) or "starting Ramadan after 29 Shaʿban when the horizon is obscured" (as Ibn ʿUmar and Ahmad contend), could lead to several practical difficulties, such as potentially observing Ramadan for either twenty-eight or thirty-one days.

SOME VERY PRACTICAL CHALLENGES

Shafat Ahmad, who has analyzed these difficulties thoroughly, contends:

> The words *fa aqduru lāhu* were meant to say what they say: estimate the duration. The actual method of estimation was left unspecified, since that would depend on the available information and analytical tools, which can change from place to place and time to time. However, people tried to make the phrase more specific and establish a simple rule applicable in all situations. One simple way to do that would be to give to the month a particular number of days in case of obscurity – 29 or 30. This raises the question of whether the same number will apply to both Shaʿban and Ramadan. The following four answers were possible depending on whether in case of obscurity both Shaʿban and Ramadan are taken to consist of 29 days or 30 days or one of them is taken to consist of 29 days and the other of 30 days:
>
> a) If there is obscurity on the 29th of Shaʿban, take that month to be 30 days and the same is true of Ramadan.
>
> In this case, you would never fast more than 30 days but sometimes you will fast only 28 days. For, suppose that both Shaʿban and Ramadan are 29 days but it is cloudy on 29th of Shaʿban and clear on 29th of Ramadan. You will count Shaʿban as 30 days and in this way miss one day of Ramadan. But if the sky is clear on the 29th of Ramadan you will be able to see the *hilāl* of Ramadan and therefore end fasting, even though you fasted only for 28 days. In places like the Caribbean islands, Trinidad, and Guyana where it is cloudy very often this process could lead even to less than 28 days of fasting.
>
> b) If there is obscurity on the 29th of Shaʿban, take that month to be 29 days and the same is true of Ramadan.
>
> In this case, you would never fast for less than 29 days but sometimes you would fast 31 or more days or have ʿId al-Fitr in Ramadan. For suppose that both Shaʿban and Ramadan are 30 days and it is cloudy on 29th of Shaʿban and clear on the 29th of Ramadan. By the rule of restricting the month to 29 days in case of obscurity, you will count Shaʿban as 29 days and thus fast on the last day of Shaʿban, but since the sky is clear on the 29th of Ramadan you will know that Ramadan has not ended. So you will fast 30 days of Ramadan and one day of Shaʿban, a total of 31 days. In case it is cloudy for several months leading to Ramadan you will need to fast even more than 31 days.
>
> c) If there is obscurity on the 29th of Shaʿban, take that month to be 30 days but if there is obscurity on the 29th of Ramadan take it to be 29 days.

> In this case you will never fast for more than 30 days but sometimes you will fast 28 days.
>
> d) If there is obscurity on the 29th of Shaᶜban, take that month to be 29 days but if there is obscurity on the 29th of Ramadan take it to be 30 days.
>
> In this case you will never fast for less than 29 days but sometimes you will fast 31 days.[39]

He makes a significant observation in his conclusion:

> Today we all assume (a), that is, in case of obscurity we should take the month as of 30 days whether it is Shaᶜban or Ramadan. It would therefore surprise some readers to hear that all of the above views have been held by Muslims. Indeed, the differences in the various narrations of the hadith about starting/ending Ramadan can be explained as attempts to reflect these interpretations."[40]

WEAKNESS OF THE IJMAᶜ (CONSENSUS) ARGUMENT

Despite this overwhelming majority's total rejection of calculations, there have always been voices of dissent among the three schools of thought, except for the Hanbali (as will be seen below). So the Hanafi, Maliki, and Shafiᶜi authorities have argued against the total rejection of calculations to establish the beginning of Ramadan and other events. Only the Hanbalis, especially in the classical period, seem to have reached a kind of consensus on this issue.

A tiny minority of earlier jurists, not to mention an ever-increasing number of contemporary jurists, disagree with such a complete dismissal. They, in opposition to the established opinion, argue that calculations are a definitive way of knowing the movements of celestial bodies and more reliable than just naked-eye sighting. Stating that neither the Qur'an nor the Sunnah clearly prohibit the use of calculations in matters of religion, they in fact find support for their arguments both in these two sources and in scientifically logical arguments.

There are three main groups among these scholars. The first one accepts calculations only in negating the beginning or end of the month. In other words, if the calculations prove that the new moon cannot be sighted, cannot have been born, and so on, then they reject all claims of naked-eye sighting, even if the witnesses are trustworthy righteous Muslims. This ancient trend can be traced back all the way to al-Subki's times. Even some of the contemporary scholars, who otherwise despise calculations, wittingly or unwittingly subscribe to this trend.

The second trend championed by Mutarrif ibn Abd ᶜAllah, ᶜUmar ibn Surayj, and al-Subki permitted the use of calculations in the case of obscurities. The third group accepts calculations in establishing the beginning of the month and also in negating or dismissing of any naked-eye sighting claims if the calculations prove otherwise. This means that Ramadan and other Islamic months can be confirmed or negated solely based on calculations. This is a recent phenomenon among such contemporary scholars as Muhammad Mustafa al-Maraghi, Ahmad M. Shakir, Mustafa al-Zarqa, ᶜAli al-Tantawi, Sharaf al-Quda, and many others. Therefore, the claims of ijmaᶜ (consensus) that naked-eye sighting or completion are the only two methods accepted by the entire Ummah are not based upon historical fact.

Moreover, there is no consensus among the majority (*al-jumhūr*) about the exact nature of naked-eye sighting, whether it is established by the sighting of one or more witnesses or a multitude of people. There also exist numerous opinions about the criteria and characteristics of these witnesses: whether they are male or female, slave or free. Nor is there any consensus on the number of witnesses needed to confirm Ramadan and Shawwal.[41] For instance, Hanafi jurists require the sighting to be witnessed by a large number of individuals when the horizon is free from obscurities. They accept the witness of one trustworthy Muslim if the sky is cloudy only to confirm Ramadan.[42]

Maliki jurists require many witnesses when the horizon is free from obscurities (e.g., such as cloud, dust or fog) and two or more trustworthy Muslims if it is cloudy. Unlike Hanafi jurists, they do not accept one witness to confirm Ramadan or Shawwal. Shafiᶜis jurists accept one trustworthy Muslim's witness in both cases, whether for Ramadan or Shawwal. Hanbali jurists accept one trustworthy witness in confirming the month of Ramadan, but require two witnesses for Shawwal.[43]

Space does not permit me to go into the details of the issues connected with the methodology of sighting. It is sufficient to note that despite apparent claims of consensus about the naked-eye sighting being the only way to confirm the month before 30 Shaᶜban, there is a marked difference among jurists in the details related to the same subject. Therefore, naked-eye sighting cannot be called a categorically absolute rule, even where there is no difference of opinion. The best description is that it is a *ẓannī* (presumptive) instead of a *qaṭᶜī* (categorical) matter according to the Shariᶜah. The Shariᶜah definitely prefers the categorical over the presumptive.

Rashid Rida observes:

وَمِنْ قَوَاعِدِ الشَّرِيعَةِ الْمُتَّفَقِ عَلَيْهَا أَنَّ الْعِلْمَ مُقَدَّمٌ عَلَى الظَّنِّ ، فَلا يُعْمَلُ بِالظَّنِّ مَعَ إِمْكَانِ الْعِلْمِ ، فَمَنْ
أَمْكَنَهُ رُؤْيَةُ الْكَعْبَةِ لا يَجُوزُ لَهُ أَنْ يَجْتَهِدَ فِي التَّوَجُّهِ إِلَيْهَا وَيَعْمَلَ بِظَنِّهِ الَّذِي يُؤَدِّيهِ إِلَيْهِ الاجْتِهَادُ.[44]

The agreed-upon legal (*sharʿī*) rules require one to prefer categorical knowledge over probable [knowledge]. It is not permitted to apply probable [knowledge] in the presence of categorical knowledge. It is not allowed for a person who can attain precise knowledge of the Kaʿbah's direction by actually sighting it to go by [a] mere estimated direction and probability.

Weakness of the Jewish Argument

One of the main reasons for rejecting a calculation-based calendar, in the view of many Muslim jurists, is to oppose the Jews' adoption of such a calendar. Many classical and modern Islamic scholars quote the prophetic narration that encourages Muslims not to imitate them, but rather to oppose them in many of their religious customs and rituals. The Jewish community had reportedly adopted the calculated calendar since the fourth century AC. Therefore, scholars like Ibn Taymiyyah and many others argue that accepting astronomical calculations as the basis of the Islamic calendar would mean imitating the Jews in their innovation and misguidance. Modern jurists, among them Abdullah Saleem,[45] contend that the Prophet was aware of this innovation and specifically ordered the Muslims not to follow that path: "We are an unlettered nation. We neither write nor calculate." This was a direct reference to the Jewish calendar and calculations.

It is useful here to analyze and discuss briefly the Jewish calendar and its history to dispel the misconception that using calculations to confirm the Islamic months would constitute a sharp deviation from the prophetic Sunnah and an imitation of the Jews in their changing the religion of Allah.

The Biblical month is a lunar month (Exodus 12:2). The Hebrews had followed the moon's movements to determine their months and festivals since Antiquity. The earlier Temple had required human witnesses to physically sight the moon in order to confirm the new month. If no witness was brought on its twenty-ninth day, the month, upon its completion of thirty days, was declared complete. The Talmud states:

> The commencement of the month was dated from the time when the earliest visible appearance of the new moon was reported to the Sanhedrin. If this happened on the 30th day of the current month, that month was considered to have

ended on the preceding 29th day, and was called deficient. But if no announcement was made on the 30th day, that day was reckoned to the current month, which was then called full, and the ensuing day was considered the first of the next month.[46]

The Mishna and the Talmud, the Jewish jurisprudential sources, emphasize the rule of naked-eye sighting:

> But if it is always defective, why should they profane it? *Because it is a religious duty to sanctify [the New Moon] on the strength of actual observation.* According to another version, R. Nahman said: We also have learnt: 'For the fixing of two New Moons the Sabbath may be profaned, for those of Nisan and of Tishri'. Now if you say that the Adar which precedes Nisan is always defective, there is no difficulty; the reason why the Sabbath may be profaned is because it is a religious duty to sanctify [the New Moon] on the strength of actual observation. But if you say that it is sometimes full and sometimes defective, why should [the Sabbath] be profaned? Let us prolong [the month] today and sanctify [the New Moon] tomorrow? If the thirtieth day happens to be on Sabbath, that is actually what we do. Here, however, we are dealing with the case where the thirty-first day happens to fall on the Sabbath [and we allow the Sabbath to be profaned because] *it is a religious duty to sanctify on the strength of actual observation.*"[47]

According to Jewish law, the Sabbath is so sacred that profaning it carries the death penalty. The Bible reports:

> And the Lord spake unto Moses, saying, "Speak thou also unto the children of Israel, saying, 'Verily my sabbaths ye shall keep: for it *is* a sign between me and you throughout your generations; that *ye* may know that I *am* the Lord that doth sanctify you. Ye shall keep the sabbath therefore; for it *is* holy unto you: every one that defileth it shall surely be put to death: for whosoever doeth *any* work therein, that soul shall be cut off from among his people. Six days may work be done; but in the seventh *is* the sabbath of rest... whosoever doeth *any* work in the sabbath day, he shall surely be put to death. Wherefore the children of Israel shall keep the sabbath, to observe the sabbath throughout their generations, *for* a perpetual covenant. It *is* a sign between me and the children of Israel for ever: for *in* six days the Lord made heaven and earth, and on the seventh day he rested, and was refreshed.'" (Exodus 31:12–17)

Yet it could be violated in order to witness the new moon's actual sighting. There was even a special Jewish rabbinical court that used to verify these witnesses and announce the new month's beginning. Rabbi Gamaliel II (80–116 AC) used to receive the reports of the witnesses in person. Afterwards, the rabbis started to use calculations to negate the months. The later portion of Talmud reports:

When R. Zera went up [to Palestine], he sent back word to them [in Babylon]: It is necessary that there should be [on New Moon] a night and a day of the new moon. This is what Abba the father of R. Simlai meant: "We calculate [according to] the new moon's birth. If it is born before midday, then certainly it will have been seen shortly before sunset. If it was not born before midday, certainly it will not have been seen shortly before sunset.' What is the practical value of this remark? — R. Ashi said: 'To [help us in] confuting the witnesses.' R. Zera said in the name of R. Nahman: 'The moon is invisible for twenty-four hours [round about new moon]. For us [in Babylon] six of these belong to the old moon and eighteen to the new; for them [in Palestine] six to the new and eighteen to the old.' What is the practical value of this remark? – R. Ashi said: 'To confute the witnesses.'"[48] [Editor's Note: "R." signifies "Rabbi."]

Rashi, the famous classical Jewish authority on Biblical and Talmudic exegesis, has explained these verses as follows:

(6) Because if the conjunction is calculated to have been after midday and they claim to have seen the new moon before nightfall, they are not telling the truth.

(7) Which would imply that in Babylon the new moon is not visible till eighteen hours after its birth.

(8) Which would imply that in Palestine the new moon is visible six hours after its birth.[49]

According to the *Jewish Encyclopedia,* testimony later gave way to mere calculations, though not without controversy:

Under the patriarchate of Rabbi Judah III (300-330), the testimony of the witnesses with regard to the appearance of the new moon was received as a mere formality, the settlement of the day depending entirely on calculation. This innovation seems to have been viewed with disfavor by some members of the Sanhedrin, particularly Rabbi Jose, who wrote to both the Babylonian and the Alexandrian communities, advising them to follow the customs of their fathers and continue to celebrate two days, an advice which was followed, and is still followed, by the majority of Jews living outside of Palestine.[50]

Two practical problems demanded their dependence upon calculations instead of naked-eye sighting. The first one was that the Old Testament connected the Jewish festivals and holidays with certain crops and seasons. Sometimes when the lunar dates of the holidays fell in the wrong season, the requisite crops and fruits were not ready. The rabbis were thus forced to introduce intercalation. The *Jewish Encyclopedia* explains:

It thus seems plain that the Jewish year was not a simple lunar year; for while the Jewish festivals no doubt were fixed on given days of lunar months, they also had a dependence on the position of the sun. Thus the Passover Feast was to be celebrated in the month of the wheat harvest..., and the Feast of Tabernacles ... took place in the fall. Sometimes the feasts are mentioned as taking place in certain lunar months (Lev. xxiii.; Num. xxviii., xxix.), and at other times they are fixed in accordance with certain crops; that is, with the solar year.[51]

Exegetes of the Talmud report the reasons for an intercalation:

The solar year, which consists of three hundred and sixty-five and a quarter days, is divided into four equal parts, each period consisting of ninety-one days and seven and a half hours. These are called respectively the Nisan (vernal), Tammuz (summer), Tishri (autumnal), Tebeth (winter) Tekufoth. The lunar year which forms the basis of our calendar comprises altogether three hundred and fifty-four days. Though according to Biblical tradition our months are to be lunar (cf. Ex. XII, 2), yet our Festivals are to be observed at certain agricultural seasons; Passover and Pentecost in the Spring; Tabernacles, or Feast of Ingathering, in the autumn. In order to harmonise the lunar and solar years, a second Adar is intercalated once in two or three years. Our text lays down certain principles by which the Intercalators are to be guided.[52]

In the later periods of Jewry, calculations were used not to determine or negate the new month as it relates to the new moon, but as it relates to the seasons in which these holidays must fall. The Talmud makes this point clear, as modern exegetes explain:

The Jewish year consists ordinarily of twelve lunar months (v. n. 5). In order to prevent the festivals from falling in the wrong seasons, it was necessary periodically to adjust the lunar calendar to the solar year: this was achieved by introducing an intercalary month (Adar II) between Adar and Nisan.[53]

Thus the Jewish calendar became luni-solar instead of a lunar one solely dependent upon the new moon's birth or sighting. The Talmud states:

The Jewish Calendar, while being lunar, takes cognisance of the solar system to which it is adjusted at the end of every cycle of nineteen years. For ritual purposes the four Tekufoth seasons are calculated according to the solar system, each being equal to one fourth of 365 days, viz. 91 days, 71/2 hours. Tekufah of Nisan (Vernal equinox) begins March 21; Tekufah of Tammuz (Summer Solstice), June 21; Tekufah of Tishri (Autumnal equinox), September 23; Tekufah of Tebeth (Winter Solstice), December 22. Should the Tekufah of Tammuz extend till after the Succoth Festival, or the Tekufah of Tebeth till the sixteenth of Nisan, the year would be intercalated, so that the festivals might fall in their due seasons, viz., Passover in Spring, Succoth in Autumn.[54]

The Sanhedrin also laid down the rules on when intercalation is and is not permitted.

> Our Rabbis taught: A year may not be intercalated except where it is necessary either for [the improvement of] roads or for [the repair of] bridges, or for the [drying of the] ovens [required for the roasting] of the paschal lambs, or for the sake of pilgrims from distant lands who have left their homes and could not otherwise reach [Jerusalem] in time. But no intercalation may take place because of [heavy] snows or cold weather or for the sake of Jewish exiles [from a distance] who have not yet set out. Our Rabbis taught: The year may not be intercalated on the ground that the kids or the lambs or the doves are too young. But we consider each of these circumstances as an auxiliary reason for intercalation.[55]

This body gave the following three reasons for the intercalation:

> Our Rabbis taught: A year may be intercalated on three grounds: on account of the premature state of the corn-crops; or that of the fruit-trees; or on account of the lateness of the Tekufah. Any two of these reasons can justify intercalation, but not one alone. All, however, are glad when the state of the spring-crop is one of them.[56]

It seems obvious that the Jewish process of calculation and intercalation is arbitrary, for it gives more significance to holidays, crops, fruits, seasons, and many similar external factors than to the actual new moon. Such an attitude contradicts the Islamic calendar, which is based on the new moon itself. Consequently, using calculation to fix the Jewish calendar is quite different from fixing the Islamic calendar based on those calculations that determine the new moon's actual birth. The Talmud states that the Jewish calendar is independent of that factor:

> The average year has six months of thirty days each, and six of twenty-nine days each. For there are about twenty-nine and one half days between one new moon and the other, whence a month of thirty days, to restore the balance, must be followed by one of twenty-nine days. However, there are more than twenty-nine and one half days between one new moon and the other, approximately twenty-nine days, twelve hours and forty minutes; furthermore, there are other causes influencing the fixing of the calendar, as the result of which the arrangement of six full and defective months undergoes certain variations, so that one year might have a larger number of full, the other more than the half of defective months. In the time of the Mishnah the Sanhedrin decreed the beginning of the new months on the basis of the testimony of witnesses who had actually seen the new moon. But even then conditions would arise (such as non-visibility of the new moon, due to cloudy weather) when the Sanhedrin would be guided by its own astronomical calculations. For such a decree the principle was adopted that no year may have more than eight, nor less than four full months.[57]

There was another problem. The civil calendars were fixed by local civil authorities who were quite often very intolerant of the Jewish community. At times, the conflict between Jewish and civil holidays led to the local authorities' persecution of the Jews. Therefore, intercalation was introduced to avoid conflict with the civil calendar and its painful consequences. The *Jewish Encyclopedia* reports:

> Under the reign of Constantius (337-361) the persecutions of the Jews reached such a height that all religious exercises, including the computation of the calendar, were forbidden under pain of severe punishment.[58]

Consequently, Rabbi Hillel II published the rules of calendar computation as well as a fixed Jewish calendar in 359 and modified it in 363, four years later. Jewish communities all over the world still use it.

Tracy R. Rich summarizes how the Jewish calendar is computed.

> The Jewish calendar is based on three astronomical phenomena: the rotation of the Earth about its axis (a day); the revolution of the moon about the Earth (a month); and the revolution of the Earth about the sun (a year). These three phenomena are independent of each other, so there is no direct correlation between them. On average, the moon revolves around the Earth in about 29 1/2 days. The Earth revolves around the sun in about 365 1/2 days, that is, about 12 lunar months and 11 days.
>
> To coordinate these three phenomena, and to accommodate certain ritual requirements, the Jewish calendar consists of 12 or 13 months of 29 or 30 days, and can be 353, 354, 355, 383, 384 or 385 days long. The linchpin of the calendar is the new moon, referred to in Hebrew as the *molad.*
>
> A new month on the Jewish calendar begins with the *molad* (pronounced moh-LAHD). *Molad* is a Hebrew word meaning "birth," and refers to what we call the "new moon" in English. The molad for the month of *Tishri* (the month that starts with *Rosh Hashanah*) is the most important one for calendar calculations, and is referred to as Molad Tishri.
>
> Note that the calculated *molad* does not necessarily correspond precisely to the astronomical new moon. The length of time from one astronomical new moon to the next varies somewhat because of the eccentric orbits of the Earth and moon; however, the *moladot* of Rabbi Hillel's calendar are set using a fixed average length of time: 29 days, 12 hours, and 793 parts (or in Hebrew, *chalakim*). The amount of time is commonly written in an abbreviated form: 29d 12h 793p.[59]

Rich also explains the practical steps required to calculate its exact dates and months:

> 1. Start with a known *molad* (and the corresponding Gregorian date, if you wish to convert your resulting date to Gregorian).
>
> 2. Determine the number of months between the known *molad* and Tishri of the year of the date you are calculating.

3. Multiply the number of months by the length of the *molad*: 29d 12h 793p.

4. Add the result to the known starting *molad*.

5. Apply the *dechiyot* (rules of postponement) to determine the date of Rosh Hashanah for the year of your date.

6. To get the Gregorian date, add the number of days elapsed calculated above to the Gregorian starting date.[60]

The Jews start their calendar with the supposed date for the beginning of creation, as reported by the Hebrew Bible. We are now in the Jewish year 5768 (2008). Computing its years, months, and days is quite complicated, and the procedure requires some mathematical calculations rather than just knowledge of astronomical calculations. That might have been why the Prophet said: "We are an unlettered people; we neither write nor compute." This reference might have been to the sophisticated procedure of calculating the Jewish new month and the year, as seen above. The Prophet would not have depended upon the Jews and their means of calculation to establish the Islamic months. That is why he simplified the procedure by asking his followers to start the month by naked-eye sighting, for they were not well versed in mathematical calculations. Moreover, he wanted to connect the new month's commencement with the new moon's birth or sighting instead of with the crops and seasons, as seems to be the case with the Jewish calendar. He eliminated all arbitrary interference with dates and so on due to external factors and wanted them to be determined by the moon, for this method would ensure that acts of worship would fall in their proper (viz., divine) time rather than at a variable time superficially calculated by humans.

إن عدة الشهور عند الله اثنا عشر شهرا في كتاب الله يوم خلق السماوات والأرض منها أربعة حرم ذلك الدين القيم فلا تظلموا فيهن أنفسكم وقاتلوا المشركين كافة كما يقاتلونكم كافة واعلموا أن الله مع المتقين. إنما النسيء زيادة في الكفر يضل به الذين كفروا يحلونه عاما ويحرمونه عاما ليواطئوا عدة ما حرم الله فيحلوا ما حرم الله زين لهم سوء أعمالهم والله لا يهدي القوم الكافرين. (التوبة: 36-37)

Behold, the number of months, in the sight of God, is twelve months, [laid down] in God's decree on the day when He created the heavens and the earth; [and] out of these, four are sacred: this is the ever true law of God. Do not, then, sin against yourselves with regard to these [months]... The intercalation [of months] is but one more instance of [their] refusal to acknowledge the truth – [a means] by which

> those who are bent on denying the truth are led astray. They declare this [intercalation] to be permissible in one year and forbidden in [another] year, in order to conform [outwardly] to the number of months which God has made sacred: and thus they make lawful what God has forbidden. Goodly seems unto them the evil of their own doings, since God does not grace with His guidance people who refuse to acknowledge the truth. (9:36–37)

These verses refer to the pre-Islamic Arabs' arbitrary intercalation in the case of their months and years, as most exegetes have reported.[61]

It is obvious from the above discussion that using astronomical calculations to determine the new moon's birth or visibility does not constitute an imitation of the Jewish calendar. Although the procedure does share some elements of Jewish calculations, it is not completely identical to it. In addition, the Jewish procedure is far more complicated and includes many factors that are external to the Islamic procedure. The Islamic *mawlid* differs from the Jewish *molad*. The Islamic calendar is purely lunar, whereas the Jewish calendar is luni-solar. There is a little similarity between them, namely, seeking to determine the new moon's birth or visibility through calculations. The remaining factors are quite different. Such a degree of similarity cannot be labeled as following the Jews in their religious innovations, as some scholars hastily portray it. The same can be said about naked-eye sighting and the requirements of the human witnesses. Jewish jurisprudence had required it since Antiquity, and some Jewish sects and scholars still continue to follow this rule literally. Given this, would observing the moon with the human eye, as many classical and modern Islamic scholars require, constitute a Jewish imitation that the Shari^cah would also forbid? I am sure the answer will be no!

Weakness of the Hardship Argument

In regard to the hardship argument, it must be noted that we now live in a "global village." In this age of instant communication, news travels all over the world within seconds. Therefore, the argument of hardship leveled by al-Nawawi and others loses its ground. In reality, it is the other way round as al-Qaradawi rightly contends.[62] Muslims all over the world, especially in the West, suffer much hardship owing to uncertainties connected with actual sighting. Some of them wait until midnight just to start their *tarāwīḥ* prayers or decide about their ᶜId prayers. There are many hardships for the Muslim employees and students. Therefore, depending upon naked-eye sighting rather than calculation has now become a source of hardship.

Perhaps these are the reasons why al-Zarqa is amazed that a large number of present-day conservative jurists adamantly reject calculations in confirming or negating Ramadan, for they use the same calculations for acts of worship that are far more important in significance as well as frequency, such as the daily prayers. Classical jurists were correct in opposing these calculations, for the science at that time had not reached the levels of authenticity and certainty that it enjoys today. They could have not based important acts of worship such as the fasting Ramadan upon calculations that were not totally precise. Are we going to drag their opposition forward at a time when the reason for their view is no longer valid? The cause and effect always go hand in hand. If the cause is no longer present, then the effect must also cease to exist.

CHAPTER 6

ARGUMENTS FOR ASTRONOMICAL CALCULATIONS

This group of scholars argues that calculations are a definitive way of knowing the movements of celestial bodies and are more accurate than just naked-eye sighting. Neither the Qur'an nor the Sunnah bans the use of calculations in matters of religion, as elaborated above. In fact, the Qur'an clearly states that the sun and the moon have precisely calculated orbits and follow them meticulously to the second: "The sun and the moon follow courses [exactly] computed" (الشمس والقمر بحسبان (الرحمن: 5.

THE MOON: THE DIVINE SOURCE OF PRECISE CALCULATIONS

The Qur'an states:

(والقمر قدرناه منازل حتى عاد كالعرجون القديم لا الشمس ينبغي لها أن تدرك القمر ولا الليل سابق النهار وكل في فلك يسبحون.) (يس: 39-40)

And the moon: We have measured for it mansions [to traverse] till it returns like the old [and withered] lower part of a date-stalk. It is not permitted for the sun to catch up with the moon, nor can the night outstrip the day: each [just] swims along in [its own] orbit [according to the law]. (36:39-40)

It also reveals that Allah created specified orbits for both of them so that people can know the number of years and the calculations.

(هو الذي جعل الشمس ضياء والقمر نورا وقدره منازل لتعلموا عدد السنين والحساب.) (يونس: 5)

It is He who made the sun to be a shining glory and the moon to be a light [of beauty], and measured out stages for it; that you might know the number of years and the calculation [of time]. (10:5)

The theme "you may know the number of the years and the calculation [of time]" occurs in 17:12 as well.

In addition, they argue that the Prophet prescribed naked-eye sighting to confirm Ramadan, for it was the only available method that ensured certainty. Sighting the new moon is not *ʿibādah* in itself, but only a means to achieve the goal of certainty. If this goal could be achieved through a different and more accurate method, then following it would be just as Islamic as naked-eye sighting. They believe that calculations are currently more precise than naked-eye sighting. Therefore, the Islamic months should be confirmed by the former method, not the latter.

Those who permit the use of calculations cite the following prophetic narrations to prove this point:

حدّثنا عبيدُاللَّهِ بنُ عبدِالمجيدِ، ثَنا مالكٌ عنْ نافعٍ، عن ابنِ عمرَ، أنَّ رسولَ اللَّهِ صلى الله عليه وسلم ذَكَرَ رمضانَ فقالَ: (لَا تصوموا حتَّى تَروا الهلالَ ولا تفطِروا حتَّى تَرَوْهُ، فإن غُمَّ عليكم فأَقدروا له).[1]

The Prophet mentioned Ramadan and said, "Do not fast until you see the moon and do not break fast until you see it. If it is cloudy, then estimate it."

حدّثنا سليمانُ بنُ حربٍ، ثَنا حمادُ بنُ زيدٍ، عنْ أيوبَ عنْ نافعٍ، عن ابنِ عمرَ، قالَ: قالَ رسولُ اللَّهِ صلى الله عليه وسلم: (إنما الشهرُ تسعٌ وعشرونَ فلا تصوموا حتى ترَوْهُ ولا تفطِروا حتى ترَوْهُ، فإن غمَّ عليكُم فأقدروا له).[2]

The month [sometimes] consists of twenty-nine days. Therefore, do not fast until you see it and do not break the fast until you see it. Calculate it if it is cloudy.

حَدَّثنا إسماعيلُ بنُ جَعْفرٍ، قال: وأخبرني عبدُ اللَّهِ بنُ دينارٍ أَنَّه سَمِعَ ابنَ عُمَرَ قالَ: قَالَ رَسُولُ اللَّهِ: (لا تَصُوموا حَتَّى تَرَوُا الهِلالَ ولا تُفْطِرُوا حَتَّى تَرَوْهُ إلا أنْ يُغَمَّ عَلَيْكُمْ فإنْ غُمَّ عَلَيْكُمْ فاقْدُرُوا لَهُ).[3]

Do not fast until you see it and do not break the fast until you see it, unless it [the sky] is cloudy. Calculate it if it is cloudy.

THE THREE ACCEPTED INTERPRETATIONS OF THESE HADITHS

Al-Nawawi states that jurists have given the following three interpretations:

وَاخْتَلَفَ الْعُلَمَاءُ فِي مَعْنَى قَوْلِهِ صلى الله عليه وسلم: (فَإِنْ غُمَّ عَلَيْكُمْ فَاقْدُرُوا لَهُ) فَقَالَ أَحْمَدُ بْنُ حَنْبَلٍ وَطَائِفَةٌ قَلِيلَةٌ: مَعْنَاهُ ضَيِّقُوا لَهُ وَقَدِّرُوهُ تَحْتَ السَّحَابِ ، وَأَوْجَبَ هَؤُلاءِ صِيَامَ لَيْلَةَ الْغَيْمِ.

The scholars differed about the prophetic report's meanings "estimate it if it is obscured from you." Ahmad ibn Hanbal and a minority group maintained that it meant restricting it and considering it under the clouds. This group mandated fasting of the day whose night was cloudy.

Imam Ahmad interprets this hadith as a command to start Ramadan on the twenty-ninth day when it is cloudy, as discussed above.

وَقَالَ مُطَرِّفُ بْنُ عَبْدِ اللَّهِ وَأَبو الْعَبَّاسِ بْنُ سُرَيْجٍ وَابْنُ قُتَيْبَةَ وَآخَرُونَ: مَعْنَاهُ قَدِّرُوهُ بِحِسَابِ الْمَنَازِلِ.

Mutarrif ibn ʿAbd Allah, Ibn Surayj, Ibn Qutaybah, and others said that it meant "Calculate it in accordance with moon phase calculations."

وَقَالَ مَالِكٌ وَأَبو حَنِيفَةَ وَالشَّافِعِيُّ وَجُمْهُورُ السَّلَفِ وَالْخَلَفِ: مَعْنَاهُ قَدِّرُوا لَهُ تَمَامَ الْعَدَدِ ثَلاثِينَ يَوْمًا.[4]

Malik, Abu Hanifah, Shafiʿi, and earlier and later *jumhūr* maintained that it meant completing thirty days.

Al-Mawsuʿah al-Fiqhiyyah states:

تَضَمَّنَ هَذَا الرَّأْيُ الْقَوْلَ بِتَقْدِيرِ الْهِلالِ بِالْحِسَابِ الْفَلَكِيِّ وَنُسِبَ إِلَى مُطَرِّفِ بْنِ عَبْدِ اللَّهِ بْنِ الشِّخِّيرِ مِنْ التَّابِعِينَ وَأَبِي الْعَبَّاسِ بْنِ سُرَيْجٍ مِنْ الشَّافِعِيَّةِ وَابْنِ قُتَيْبَةَ مِنْ الْمُحَدِّثِينَ. وَقَالَ ابْنُ عَبْدِ الْبَرِّ: لا يَصِحُّ عَنْ مُطَرِّفٍ، وَنَفَى نِسْبَةَ مَا عُرِفَ عَنْ ابْنِ سُرَيْجٍ إِلَى الشَّافِعِيِّ لانَّ الْمَعْرُوفَ عَنْهُ مَا عَلَيْهِ الْجُمْهُورُ. وَنَقَلَ ابْنُ رُشْدٍ عَنْ مُطَرِّفٍ قَوْلَهُ: يُعْتَبَرُ الْهِلالُ إِذَا غُمَّ بِالنُّجُومِ وَمَنَازِلِ الْقَمَرِ وَطَرِيقِ الْحِسَابِ ، قَالَ: وَرُوِيَ مِثْلُ ذَلِكَ عَنْ الشَّافِعِيِّ فِي رِوَايَةٍ ، وَالْمَعْرُوفُ لَهُ الْمَشْهُورُ عَنْهُ أَنَّهُ لا يُصَامُ إِلا بِرُؤْيَةٍ فَاشِيَةٍ أَوْ شَهَادَةٍ عَادِلَةٍ كَالَّذِي عَلَيْهِ الْجُمْهُورُ.[5]

This opinion holds astronomical calculations to be a sound method of estimating the moon's phases. It has been attributed to Mutarrif ibn ʿAbd Allah ibn al-Shikhir (a Successor), Abu al-ʿAbbas ibn Sarayj (a Shafiʿi), and Ibn Qutaybah (a hadith scholar). Ibn ʿAbd al-Barr denied that Mutarrif espoused that view and rejected what Ibn Sarayj attributed to Shafiʿi because it was known that the latter maintained the majority (*jumhūr*) opinion. Ibn Rushd narrated Mutarrif's statement that astronomical calculations can determine the new moon in the case of obscurities and narrated that this view is attributed to Shafiʿi in one of the reports. Shafiʿi's known opinion is that fasting may be observed only with naked-eye sighting or by the witness of a trustworthy Muslim, as the majority of jurists contend.

Linguistically and contextually, the word in this hadith leads to the meaning of التَّقْدِير, as Abu Sulayman Ahmad ibn Muhammad ibn Ibrahim al-Khattabi prefers, that is, it gives the sense of counting and calculation in the case of cloudy weather or the lack of visibility. That is why scholars like al-Khattabi, al-Dawudi, and many others take it to mean that if it happens to be cloudy on 29 Shaʿban, then relying on the calculations is not only permitted, but is actually required, by the Sunnah.[6]

Al-Baji reports that Abu ʿAbd Allah Muhammad ibn Saʿid al-Dawudi al-Zahiri has interpreted the hadith in this way:

وَذَكَرَ الدَّاوُدِيُّ أَنَّهُ قِيلَ فِي مَعْنَى قَوْلِهِ فَاقْدُرُوا لَهُ أَيْ قَدِّرُوا الْمَنَازِلَ.[7]

Al-Dawudi stated that the prophetic statement "estimate (calculate) it" has been interpreted to mean "calculating the moon's phases."

Ibn Daqiq al-ʿId reports that some Maliki scholars from Baghdad and some leading Shafiʿi authorities adopted this position, especially in regard to the astronomer himself, who is required to start fasting on the day his calculations determine it to be the first day of Ramadan.[8] He also reports that Mutarrif ibn Shikhir maintained that the astronomer must follow his calculations in confirming Ramadan. Abu al-ʿAbbas ibn Surayj, a Shafiʿi, states that the meaning of "calculate" is addressed to people who have the knowledge of calculation, whereas "sighting" is a method for ordinary Muslims.[9]

Al-Qarrafi narrates that the Malikis permit the use of calculation to determine Ramadan.

وَذَكَرَ الْقَرَافِيُّ قَوْلا آخَرَ لِلْمَالِكِيَّةِ بِجَوَازِ اعْتِمَادِ الْحِسَابِ فِي إِثْبَاتِ الاهْلِيَّةِ.[10]

Al-Qarrafi has reported another ruling of Maliki jurists that accepts astronomical calculations as a valid method of confirming new moons (months).

THE HADITH OF THE DAJJAL

Although interpretating the hadith discussed above as "estimation" conflicts with the classical majority opinion, it is in line with the linguistic meaning of "فَاقْدُرُوا لَهُ." The same phrase is used in the famous hadith of the Dajjal, in which the Prophet informed his Companions that at the time of the Dajjal, real time would seem to expand so much that a day would be equal to a year, a month, or even a week. The Companions asked how they should perform the five daily prayers in such a situation. In response, the Prophet replied, "فَاقْدُرُوا لَهُ," meaning "Calculate it." There is no way to interpret the phrase as twenty-nine or thirty days or completion. It definitely means estimation. The hadith is as follows:

حدثنا صَفْوَانَ بنُ صَالِحٍ الدِّمَشْقِيُّ المُؤَذِّن أخبرنا الْوَلِيدُ أخبرنا ابنُ جَابِرٍ حدَّثني يَحْيَى بنُ جَابِرٍ الطَّائِيُّ عن عَبْدِ الرَّحْم عنِ بن جُبَيْرٍ بنِ نُفَيْرٍ عن أَبِيهِ عن النَّوَّاسِ بنِ سَمْعَانَ الْكِلَابِيِّ، قال: (ذَكَرَ رَسُولُ الله صلى الله عليه وسلم الدَّجَّالَ فقالَ: إنْ يَخْرُجْ وأَنا فِيكم حَجِيجُهُ دُونَكُم وَإِنَّ يَخْرُجْ وَلَسْتُ فِيكُم فامْرُؤٌ حَجِيجٌ نَفْسُهُ، وَالله خَلِيفَتِي عَلَى كُلِّ مُسْلِمٍ، فَمنْ أَدْرَكَهُ مِنْكم فَلْيَقْرَأْ عَلَيْهِ بَفَواتِحِ سُورَةِ الْكَهْفِ فإِنَّهَا جِوَارُكم مِنْ فِتْنَتِهِ. قُلْنا: وَمَا لَبْثُهُ في الأَرْضِ. قالَ: أَرْبَعُونَ يَوْماً، يَوْمٌ

كَسَنَةٍ، وَيَوْمٌ كَشَهْرٍ، وَيَوْمٌ كَجُمْعَةٍ، وَسَائِرُ أَيَّامِهِ كَأَيَّامِكُم. فقُلْنا: يَا رَسُولَ الله هذَا الْيَوْمُ الَّذِي كَسَنَةٍ أَتَكْفِينَا فِيهِ صَلاَةَ يَوْمٍ وَلَيْلَةٍ؟ قال: لَا، أَقْدُرُوا لَهُ قَدْرَهُ).[11]

Al-Nawwas ibn Samʿan narrated: "The Prophet mentioned the Dajjal and said: 'If he comes forth while I am among you, I shall contend with him on your behalf; but if he comes forth while I am not among you, a man must contend on his own behalf, and Allah will take care of every Muslim on my behalf [and safeguard him against his evil]. Those of you who will survive to see him should recite over him the opening verses of *Surat al-Kahf*, for that will protect him from his trial." We asked, '[O Prophet of Allah], how long will he stay on earth?' He said, 'For forty days; one day like a year, one day like a month, one day like a week, and the rest of the days will be like your days.' We asked, 'O Prophet of Allah, will one day's prayer suffice for the prayers of the day equal to one year?' Thereupon he said, 'No, but you must estimate the time [and then observe prayer].'"

ذكر رسول الله صلى الله عليه وسلم الدجال ذات غداة، فخفض فيه ورفع، حتى ظنناه في طائفة النخل، فلما رحنا إليه عرف ذلك في وجوهنا، فسألناه، فقلنا: يا رسول الله ذكرت الدجال الغداة فخفضت فيه ورفعت حتى ظنناه في طائفة النخل؟ قال: غير الدجال أخوف مني عليكم، فإن يخرج وأنا فيكم فأنا حجيجه دونكم، وإن يخرج ولست فيكم فامرؤ حجيج نفسه، والله خليفتي على كل مسلم، إنه شاب جعد، قطط، عينه طافية، وأنه يخرج خلة بين الشام والعراق، فعاث يميناً وشمالاً، يا عباد الله اثبتوا، قلنا: يا رسول الله ما لبثه في الأرض؟ قال: أربعين يوماً، يوم كسنة ويوم كشهر ويوم كجمعة وسائر أيامه كأيامكم، قلنا: يا رسول الله فذلك اليوم الذي هو كسنة أيكفينا فيه صلاة يوم وليلة؟ قال: لا، أقدروا له قدره.[12]

Al-Nawwas ibn Samʿan narrated: "The Prophet mentioned the Dajjal one day in the morning. He sometimes described him to be insignificant and sometimes described [his turmoil] as very significant [and we felt] as if he were in the clump of date palms. When we went to him [the Prophet] in the evening and he read [the signs of fear] on our faces, he asked, 'What is the matter with you?' We said, 'O Prophet of Allah, you mentioned the Dajjal this morning [sometimes describing him] to be insignificant and sometimes very important until we began to feel that he were present in the clump of date palms nearby.' Thereupon he said, 'I harbor fear in regard to you in so many other things besides the Dajjal. If he comes forth while I am among you, I shall contend with him on your behalf, but if he comes forth while I am not among you, a man must contend on his own behalf and Allah will take care of every Muslim on my behalf [and safeguard him against his turmoil]. He [the Dajjal] will be a young man with wavy, cropped hair, and a blind eye. He will appear on the road between Syria and Iraq and will spread mischief right and left. O servant of Allah, Adhere [to the path of Truth].' We asked, 'O Prophet of Allah, how long will he stay on earth?' He said, 'For forty days; one day like a year and one day like a month and one day like a week and the rest of the days will be like your days.' We asked, 'O Prophet of Allah, would one day's prayer suffice for the prayers of a day equal to one year?' Thereupon he said, 'No, but you must make an estimate of time [and then observe prayer].'"

THE "ESTIMATION" INTERPRETATION OF THE HADITH IS MORE ACCURATE

In view of these prophetic reports, the interpretation of "أقدروا له" as calculating the month or the phases of the moon is perhaps more appropriate than the other two interpretations. That is why some authorities in the Hanafi, Maliki, and Shafiʿi have no problem in accepting calculations for this purpose. There is a single report from Hammad that Ibn ʿUmar narrated from the Prophet:

حدثنا سُلَيْمانُ بنُ دَاوُدَ الْعَتَكِيُّ أخبرنا حَمَّادٌ أخبرنا أيُّوبُ عن نَافِعٍ عن ابن عُمَرَ، قال قال رَسُولُ الله صلى الله عليه وسلم: الشَّهْرُ تِسْعٌ وَعِشْرُونَ فَلَا تَصُومُوا حَتَّى تَرَوْهُ وَلَا تُفْطِرُوا حَتَّى تَرَوْهُ. فَإِنْ غُمَّ عَلَيْكُمْ فَاقْدُرُوا لَهُ ثَلَاثِينَ.[13]

Ibn Umar reports that the Prophet said, "The month is twenty-nine days. Therefore do not fast until you see it and do not break the fast until you see it. Complete thirty days if it is obscured from you."

This narration is the only report that includes "estimate for it thirty days" instead of "estimate it." It is an oddly detached report, coming as it does through only one narrator, and thus cannot be accepted against such a variety of reports from Ibn ʿUmar through Nafiʿ, what the scholars call "the golden chain." Ibn Qudamah observes:

وَرِوَايَةُ ابْنِ عُمَرَ: "فَاقْدُرُوا لَهُ ثَلَاثِينَ" مُخَالِفَةٌ لِلرِّوَايَةِ الصَّحِيحَةِ الْمُتَّفَقِ عَلَيْهَا، وَلِمَذْهَبِ ابْنِ عُمَرَ وَرَأْيِهِ.[14]

The report from Ibn ʿUmar that [says] "count it thirty" opposes the other agreed authentic narration from him. It also contradicts his opinion and *madhhab*.

Al-Subki, a Shafiʿi who has discussed the question of calculation in great detail, rejects even a trustworthy witnesses if the calculations negate the possibility of sighting the moon:

وَهَهُنَا صُورَةٌ أُخْرَى وَهُوَ أَنْ يَدُلَّ الْحِسَابُ عَلَى عَدَمِ إمْكَانِ رُؤْيَتِهِ وَيُدْرَكُ ذَلِكَ بِمُقَدَّمَاتٍ قَطْعِيَّةٍ وَيَكُونُ فِي غَايَةِ الْقُرْبِ مِنْ الشَّمْسِ فَفِي هَذِهِ الْحَالَةِ لا يُمْكِنُ فَرْضُ رُؤْيَتِنَا لَهُ حِسًّا لأَنَّهُ يَسْتَحِيلُ فَلَوْ أَخْبَرَنَا بِهِ مُخْبِرٌ وَاحِدٌ أَوْ أَكْثَرُ مِمَّنْ يَحْتَمِلُ خَبَرُهُ الْكَذِبَ أَوْ الْغَلَطَ فَالَّذِي يُتَّجَهُ قَبُولُ هَذَا

الْخَبَرِ وَحَمْلُهُ عَلَى الْكَذِبِ أَوْ الْغَلَطِ وَلَوْ شَهِدَ بِهِ شَاهِدَانِ لَمْ تُقْبَلْ شَهَادَتُهُمَا لانَّ الْحِسَابَ قَطْعِيٌّ وَالشَّهَادَةَ وَالْخَبَرَ ظَنِّيَّانِ وَالظَّنُّ لا يُعَارِضُ الْقَطْعَ فَضْلا عَنْ أَنْ يُقَدَّمَ عَلَيْهِ وَالْبَيِّنَةُ شَرْطُهَا أَنْ يَكُونَ مَا شَهِدْت بِهِ مُمْكِنًا حِسًّا وَعَقْلا وَشَرْعًا فَإِذَا فُرِضَ دَلالَةُ الْحِسَابِ قَطْعًا عَلَى عَدَمِ الامْكَانِ اسْتَحَالَ الْقَبُولُ شَرْعًا لاسْتِحَالَةِ الْمَشْهُودِ بِهِ وَالشَّرْعُ لا يَأْتِي بِالْمُسْتَحِيلاتِ.[15]

There is another scenario: if the astronomical calculations prove that a sighting is impossible and this is known from categorical inferences, such as the moon being too close to the sun at the time of sunset, then it is not possible to see it with our human senses because such a sighting is impossible. Now, if one person or two or a group of untrustworthy individuals produce a witness [claiming to have] sighted it, their witness must be rejected. [This is] because the astronomical calculations are precise, whereas the human witness and news are hypothetical, and the hypothetical cannot be accepted against something categorical, let alone given priority over it. For a witness to be accepted, it is required that what is being witnessed is possible Islamically [lawfully], logically, and sensually. Therefore, if the astronomical calculations prove that a sighting is impossible, it would be impossible to accept any claim of that Islamically, because what is being witnessed is not there and the Shariʿah does not create something self-contradictory and impossible in itself.

Al-Subki's main argument is that calculations are accurate, whereas there is the possibility of confusion or mistake in naked-eye sighting. Therefore, the Shariʿah would not prefer a probable method to a reliable and accurate one. He further argues that the Shariʿah did not require us to accept the news of a naked-eye sighting without verification. We cannot base our fasting solely on the claims of witnesses, for the Shariʿah does not ask for that. Verification of the news is essential. How often have we seen people giving false witness, either unintentionally or intentionally, for some hidden motive?[16] Thus he advises the authorities to take the calculations into consideration, especially in negating the claimed naked-eye sighting when calculations prove that such cannot be the case. He also advises them not to pay too much attention to the views that prohibit the use of calculation in matters of religion, for the Shariʿah does not forbid it.

فَيَجِبُ عَلَى الْحَاكِمِ إذَا جَرَّبَ مِثْلَ ذَلِكَ وَعَرَفَ مِنْ نَفْسِهِ أَوْ بِخَبَرِ مَنْ يَثِقُ بِهِ أَنَّ دَلالَةَ الْحِسَابِ عَلَى عَدَمِ إمْكَانِ الرُّؤْيَةِ أَنْ لا يَقْبَلَ هَذِهِ الشَّهَادَةَ وَلا يُثْبِتَ بِهَا وَلا يَحْكُمَ بِهَا، وَيُسْتَصْحَبُ الاصْلُ فِي بَقَاءِ الشَّهْرِ فَإِنَّهُ دَلِيلٌ شَرْعِيٌّ مُحَقَّقٌ حَتَّى يَتَحَقَّقَ خِلافُهُ، وَلا نَقُولُ الشَّرْعُ أَلْغَى قَوْلَ الْحِسَابِ مُطْلَقًا.[17]

It is obligatory upon the ruler to reject the witness of such people if he knows himself or from a trustworthy person that the calculations have proved that a physical sighting is impossible. He should neither accept that witness nor give any ruling based upon such a claim. The month should be considered as continuing until it is proven otherwise, as the Shariʿah requires. And we do not say that the Shariʿah has banned the use of astronomical calculations at all.

Al-Subki is careful to differentiate between precise calculations and those based upon anticipation or probability. He asks the judges to use their sense of judgment when the calculations are probable.

وَمَرَاتِبُ مَا يَقُولُهُ الْحِسَابُ فِي ذَلِكَ مُتَفَاوِتَةٌ مِنْهَا مَا يَقْطَعُونَ بِعَدَمِ إِمْكَانِ الرُّؤْيَةِ فِيهِ فَهَذَا لا رَيْبَ عِنْدَنَا فِي رَدِّ الشَّهَادَةِ بِهِ وَمِنْهَا مَا لا يَقْطَعُونَ فِيهِ بِعَدَمِ الامْكَانِ وَلَكِنْ يَسْتَعِدُّونَ فَهَذَا مَحَلُّ النَّظَرِ فِي حَالِ الشُّهُودِ وَحِدَّةِ بَصَرِهِمْ وَيَرَى أَنَّهُمْ مِنْ احْتِمَالِ الْغَلَطِ وَالْكَذِبِ يَتَفَاوَتُ ذَلِكَ تَفَاوُتًا كَبِيرًا وَمَرَاتِبَ كَثِيرَةً فَلِهَذَا يَجِبُ عَلَى الْقَاضِي الاجْتِهَادُ وُسْعَ الطَّاقَةِ.[18]

There are many types of calculation. We have no doubt in our minds that a human witness cannot be accepted against accurately precise calculations. However, when the calculations are not certain but probable, then weight should be given to the human witness and his capability of sighting, such as strength of vision etc. ... In that case, the judge must use his judgment to the best of his ability.

He concludes that calculations are more certain than naked-eye sighting and that the probability of error is greater in the second case than in the first case:

إذَا شَهِدَ عِنْدَنَا اثْنَانِ أَوْ أَكْثَرُ مِمَّنْ يَجُوزُ كَذِبُهُمَا أَوْ غَلَطُهُمَا بِرُؤْيَةِ الْهِلالِ وَقَدْ دَلَّ حِسَابُ تَسْيِيرِ مَنَازِلِ الْقَمَرِ عَلَى عَدَمِ إِمْكَانِ رُؤْيَتِهِ فِي ذَلِكَ الَّذِي قَالا: إِنَّهُمَا رَأَيَاهُ فِيهِ تُرَدُّ شَهَادَتُهُمَا لانَّ الامْكَانَ شَرْطٌ فِي الْمَشْهُورِ بِهِ وَتَجْوِيزُ الْكَذِبِ وَالْغَلَطِ عَلَى الشَّاهِدَيْنِ الْمَذْكُورَيْنِ أَوْلَى مِنْ تَجْوِيزِ انْخِرَامِ الْعَادَةِ فَالْمُسْتَحِيلُ الْعَادِيُّ وَالْمُسْتَحِيلُ الْعَقْلِيُّ لا يُقْبَلُ الاقْرَارُ بِهِ وَلا الشَّهَادَةُ فَكَذَلِكَ الْمُسْتَحِيلُ الْعَادِيُّ.[19]

If two or more untrustworthy people claim to have sighted the new moon and the astronomical calculations indicate that the sighting is not possible, then their witness is rejected because possibility is a condition for acceptability. It is more possibility that these witnesses are wrong rather than the nature [natural phenomenon i.e, calculations] being wrong. It is not accepted to approve of or accept witnesses about something which is a natural or intellectual impossibility.

Al-Subki knew that this question had not been discussed in such detail in his *madhhab* or before his time. Therefore, he felt comfortable in forcefully expressing his conclusions based upon his deep understanding of the subject.[20] In fact, he seems to have been quite ahead of his time and to have generated a heated debate over this question with some scholars. Considered a *mujtahid* (an authority in his school of thought), he concludes the discussion with the following interesting comments:

قَدْ يَحْصُلُ لِبَعْضِ الاغْمَارِ وَالْجُهَّالِ تَوَقُّفٌ فِيمَا قُلْنَاهُ وَيَسْتَنْكِرُ الرُّجُوعَ إِلَى الْحِسَابِ جُمْلَةً وَتَفْصِيلا وَيَجْمُدُ عَلَى أَنَّ كُلَّ مَا شَهِدَ بِهِ شَاهِدَانِ يَثْبُتُ، وَمَنْ كَانَ كَذَلِكَ لا خِطَابَ مَعَهُ وَنَحْنُ إِنَّمَا نَتَكَلَّمُ مَعَ مَنْ لَهُ أَدْنَى تَبَصُّرٍ وَالْجَاهِلُ لا كَلامَ مَعَهُ.[21]

Some recklessly ignorant [person] may hesitate to accept what we have stated. He might regard it as abhorrent to resort to calculation in part or in whole and may be stuck with the idea that whatever is witnessed by two persons is proven. No conversation can take place with such a rigid person. We are talking to those who at least enjoy the [knowledge of] basic logic. We cannot talk to the ignorant.

Al-Qaradawi ponders what how al-Subki would have viewed calculations and their authenticity, in regard even to acts of worship, if he had seen the scientific revolution of our time.

فكيف لو عاش السبكى إلى عصرنا هذا ورأى من تقدم علم الفلك ... كما أشرنا إلى بعضه؟[22]

What would have been the response of al-Subki had he lived during our times and witnessed all this progress in the science of astronomy ... as we have just briefly mentioned?

Other scholars such al-ʿAbbadi and Ibn Daqiq are reported to have agreed with him on this issue. Zakariyya ibn Muhammad al-Ansari reports:

لَكِنْ نَقَلَ الْقَلْيُوبِيُّ عَلَى الْجَلالِ عَنْ الْعَبَّادِيِّ أَنَّهُ قَالَ إِذَا دَلَّ الْحِسَابُ الْقَطْعِيُّ عَلَى عَدَمِ رُؤْيَتِهِ لَمْ يُقْبَلْ قَوْلُ الشُّهُودِ الْعُدُولِ بِرُؤْيَتِهِ وَتُرَدُّ شَهَادَتُهُمْ بِهَا وَلا يَجُوزُ الصَّوْمُ حِينَئِذٍ وَمُخَالَفَةُ ذَلِكَ مُعَانَدَةٌ وَمُكَابَرَةٌ.[23]

Al-ʿAbbadi said that the witness of even a trustworthy [person] would not be accepted if accurate astronomical calculations refute the possibility of a sighting. Their witnesses must be rejected owing to the calculations and fasting would not be allowed in such a case. Opposing this would be nothing short of stubbornness and haughtiness.

Ibn Hajar al-Asqalani reports that Ibn Daqiq al-ʿId said that if calculations established the fact that the moon was there and could be sighted, but that cloudy weather prevented it from being sighted, then fasting would be obligatory. This constitutes a valid Islamic reason to follow the calculations.[24]

Ibn Daqiq himself argues for this view:

وَأَمَّا إِذَا دَلَّ الْحِسَابُ عَلَى أَنَّ الْهِلالَ قَدْ طَلَعَ مِنْ الافُقِ عَلَى وَجْهٍ يُرَى، لَوْلا وُجُودُ الْمَانِعِ – كَالْغَيْمِ مِثْلا فَهَذَا يَقْتَضِي الْوُجُوبَ، لِوُجُودِ السَّبَبِ الشَّرْعِيِّ. وَلَيْسَ حَقِيقَةُ الرُّؤْيَةِ بِشَرْطٍ مِنْ اللُّزُومِ؛ لانَّ الاتِّفَاقَ عَلَى أَنَّ الْمَحْبُوسَ فِي الْمَطْمُورَةِ إِذَا عُلِمَ بِإِكْمَالِ الْعِدَّةِ، أَوْ بِالاجْتِهَادِ بِالامَارَاتِ: أَنَّ الْيَوْمَ مِنْ رَمَضَانَ، وَجبَ عَلَيْهِ الصَّوْمُ وَإِنْ لَمْ يَرَ الْهِلالَ. وَلا أَخْبَرَهُ مَنْ رَآهُ.[25]

If the calculations show that the new moon is born and can be seen over the horizon, but cannot be seen owing to obscurities such as clouds, then this makes it obligatory to fast. This constitutes an Islamic reason to confirm the month [from calculations]. The actual sighting is not a prerequisite to the fasting. There is agreement [among the jurists] that if someone was imprisoned in the basement and knew, either by completing thirty days or by diligence in following the signs, that the month of Ramadan had started, then he would be required to observe fasting even if he had neither sighted the moon himself nor had been informed by someone who actually sighted it.

Some Hanafi scholars, among them Muhammad ibn Muqatil, not only espoused the same views but even consulted astronomers and accepted their calculations regarding the lunar months.

قَالَ بَعْضُ أَصْحَابِنَا رحمهم الله لا بَأْسَ بِالاعْتِمَادِ عَلَى قَوْلِ الْمُنَجِّمِينَ وَعَنْ مُحَمَّدِ بْنِ مُقَاتِلٍ أَنَّهُ كَانَ يَسْأَلُهُمْ وَيَعْتَمِدُ عَلَى قَوْلِهِمْ بَعْدَ أَنْ يَتَّفِقَ عَلَى ذَلِكَ جَمَاعَةٌ مِنْهُمْ.[26]

Some of our scholars are of the opinion that there is nothing wrong in depending upon the astronomical calculations. Indeed, Muhammad ibn Muqatil used to inquire of astronomers about the calculations and depend upon them if the calculations were confirmed by a group of astronomers.

Ibn ʿAbidin narrates the difference of opinion among the Hanafis.

الْخِلافُ فِي جَوَازِ الاعْتِمَادِ عَلَيْهِمْ، وَقَدْ حَكَى فِي الْقُنْيَةِ الاقْوَالَ الثَّلاثَةَ فَنَقَلَ أَوَّلا عَنْ الْقَاضِي عَبْدِ الْجَبَّارِ، وَصَاحِبِ جَمْعِ الْعُلُومِ أَنَّهُ لا بَأْسَ بِالاعْتِمَادِ عَلَى قَوْلِهِمْ، وَنَقَلَ عَنْ ابْنِ مُقَاتِلٍ أَنَّهُ كَانَ يَسْأَلُهُمْ وَيَعْتَمِدُ عَلَى قَوْلِهِمْ إِذَا اتَّفَقَ عَلَيْهِ جَمَاعَةٌ مِنْهُمْ.[27]

There is a disagreement about trusting the calculations. There are three opinions narrated in al-Qunyah. Firstly, the opinion of al-Qadi ʿAbd al-Jabbar and the author of *Jamʿ al-ʿUlūm* is that there is nothing wrong in accepting the calculations. It is narrated that Ibn Muqatil used to consult the astronomers and depend upon their calculations if a group of them agreed upon it.

It is clear from this discussion that well-versed Shafiʿi, Maliki, and Hanafi authorities have opined that calculations can sometimes be used to determine the beginning and end of Ramadan. It seems that most of these jurists have supported the use of calculations in negating, rather than confirming, Ramadan. Some of them allowed the use of calculations in confirmation also. However, Ibn Daqiq al-ʿId and other jurists allowed the use of calculations to attain confirmation in the case of cloudy weather only.

MODERN JURISTS AND CALCULATIONS

Life is changing rapidly in modern times. Among the contemporary scholars, M. Mustafa al-Maraghi, ʿAli al-Tantawi, Ahmad M. Shakir, Mustafa al-Zarqa, Sharaf al-Quda, and others argue that modern science has reached a level of authenticity and precision as regards calculation that there is no longer any need for naked-eye sighting. Given that the Shariʿah required sighting when most of the Ummah was illiterate and ignorant of astronomy and other sciences relating to calculation, and since we have reached the level of certainty in these matters, we must use calculations to determine the Islamic months.

Ahmad Shakir contends that the command to depend solely upon naked-eye sighting came with a condition: the Muslims of that time did not know how to write or calculate. Hafiz ibn Hajr remarks:

والمراد أهل الإسلام الذين بحضرته عند تلك المقالة، وهو محمول على أكثرهم، أو المراد نفسه صلى الله عليه وسلم. وقيل للعرب اميون لأن الكتابة كانت فيهم عزيزة. قال الله تعالى (هو الذي بعث في الأميين رسولاً منهم) ولا يرد على ذلك أنه كان فيهم من يكتب ويحسب لأن الكتابة كانت فيهم قليلة نادرة. والمراد بالحساب هنا حساب النجوم وتسييرها و لم يكونوا يعرفون من ذلك أيضاً إلا النـــزر اليسير، فعلق الحكم بالصوم وغيره بالرؤية لرفع الحرج عنهم في معاناة حساب التسيير.[28]

The reference in the hadith is to the Muslims who were present with the Prophet when he uttered these words. It refers to the majority of them [namely, that they were illiterate] or it could be that the Prophet is referring to himself. The Arabs were called illiterate because writing skills were quite lacking among them. Allah has said, "It is He who has sent among the illiterate a messenger from among themselves." This fact cannot be refuted by the assertion that among the Arabs there were individuals who could write or calculate because the writing skills were very rare among them. In addition, the reference to calculation in the hadith is to astronomical calculation. They did not know much about astronomical calculation except a very negligible part of it. That is why the Prophet connected the ruling of fasting with the physical sighting to avoid causing any hardship to them.

In view of this, Shakir argues that it is an established rule of Islamic jurisprudence that the cause and effect go hand in hand. So, since the Ummah has become literate and started writing/calculating, the effect must be modified.

لأن الأمر باعتماد الرؤية وحدها جاء معللا بعلة منصوصة، وهى أن الأمة (أمية لا تكتب ولا تحسب)، والعلة تدور مع المعلول وجودا وعدما، فإذا خرجت الأمة عن أميتها، و صارت تكتب و تحسب، أعنى صارت فى مجموعها ممن يعرف هذه العلوم، و أمكن الناس -عامتهم و خاصتهم- أن يصلوا إلى اليقين والقطع فى حساب أول الشهر، و أمكن أن يثقوا بهذا الحساب ثقتهم بالرؤية أو أقوى... وجب أن يرجعوا إلى اليقين الثابت، و أن يأخذوا فى إثبات الأهلة بالحساب وحده، و ألا يرجعوا إلى الرؤية إلا حين يستعصى عليهم العلم به.[29]

The prophetic command asking [Muslims] to depend only upon physical sighting came also with the specified reason for doing so elaborated by the same text. The specified reason was that the Ummah of that time did not know how to write or calculate. In addition, the cause and effect always go hand in hand. Now, when the Ummah has abandoned its unlettered status and started writing and calculating – I mean that there exist a number of people among the Muslims who know these sciences – and it has become possible for all Muslims to know the precise calculations about the beginning of the month, once the Ummah can trust the accuracy of the calculations just like they trusted physical sighting or even more, then it has become obligatory upon them to follow only authentic calculations, not sighting, to confirm Ramadan. The only exception is if the calculations are hard to obtain.

He further argues that the new moon's birth begins the new month.

وإذا وجب الرجوع إلى الحساب وحده بزوال علة منعه، وجب أيضا الرجوع إلى الحساب الحقيقى للأهلة، وإطراح إمكان الرؤية و عدم إمكانها، فيكون أول الشهر الحقيقي الليلة التي يغيب فيها الهلال بعد غروب الشمس، و لو بلحظة واحدة.[30]

Now, once it has become obligatory to turn to astronomical calculation only, because the reason for its prohibition is gone, then it becomes obligatory also to turn to the accurate calculations connected with the new months and the possibility or impossibility of sighting. Therefore, the precise beginning of the new month will be the evening,when the moon is setting after sunset, even if a second after sunset.

He states that starting and ending the Islamic months with calculations rather than naked-eye sighting is the most appropriate *fiqhī* position for our time and that it conforms to the true spirit of the relevant hadith.

ولقد أرى قولى هذا أعدل الأقوال، وأقربها إلى الفقه السليم، و إلى الفهم الصحيح للأحاديث الواردة فى هذا الباب.[31]

I maintain that this statement of mine is the most fair of all. This (position) is the closest to the correct fiqh and to the appropriately genuine understandings of the hadiths reported in connection with this issue.

Mustafa al-Zarqa, after a detailed discussion of the question, concludes that there is nothing in the Shariʿah's rules that prevents contemporary Muslims from accepting such calculations.

وما دام من البديهات أن رؤية الهلال الجديد ليست فى ذاتها عبادة فى الإسلام، و إنما هي وسيلة لمعرفة الوقت، وكانت الوسيلة الوحيدة الممكنة فى أمة أمية لا تكتب ولا تحسب، و كانت أميتها هي العلة فى الأمر بالاعتماد على العين الباصرة، وذلك بنص الحديث النبوي مصدر الحكم، فما الذي يمنع شرعا أن نعتمد الحساب الفلكي اليقيني، الذي يعرفنا مسبقا بموعد حلول الشهر الجديد، ولا يمكن أن يحجب علمنا حينئذ غيم ولا ضباب إلا ضباب العقول؟[32]

It is an established fact that sighting the new moon is not an act of worship in itself. It is just a means to know the timings. It was the only means available to the unlettered nation, which did not know how to write or calculate. Its unlettered status was the sole reason for dependence upon physical sighting. This is clear from the text of the prophetic tradition that is the original source of that ruling. Islamically speaking, what prevents us now from depending upon the accurate astronomical calculations that can determine for us, quite ahead of time, the beginning of the new month? No cloud or fog can obscure our knowledge of the month then, except the fog or dust of the intellect.

After much elaboration of the subject, al-Qaradawi concludes that Islam, which prescribes naked-eye sighting as a valid method to confirm Ramadan, would definitely prefer to accept calculations as the more valid method, because there could be doubt about or mistakes in human sighting, but not in the accuracy of calculations. Therefore, accepting these calculations conforms exactly to the true spirit of the Shariʿah. The Ummah can be spared confusion and problems by following calculations.

إن الأخذ بالحساب القطعي اليوم وسيلة لإثبات الشهور، يجب أن يقبل من باب "قياس الأولى" بمعنى أن السنة التى شرعت لنا الأخذ بوسيلة أدنى، لما يحيط بها من الشك والاحتمال -وهي الرؤية- لا ترفض وسيلة أعلى وأكمل وأوفى بتحقيق المقصود، والخروج بالأمة من الاختلاف الشديد فى تحديد بداية صيامها و فطرها وأضحها... وهي وسيلة الحساب القطعي.[33]

Currently, astronomical calculations are a better means to establish the months. They must be accepted, for they are a better choice. The Sunnah, which prescribed for us the lesser method of actual sighting with all possibilities of doubt and probabilities, would never reject a perfectly superior method (calculations) that better realizes the original objective (of certainty) and will bring the Ummah out of this serious controversy, which takes place at the times of confirming the month of Ramadan, ʿId al-Fitr, and ʿId al-Adha. The precise astronomical calculations constitute that method.

Rashid Rida argues also that:

إِنَّ الَّذِينَ لَمَّ يُبِيحُوا الْعَمَلَ بِالْحِسَابِ قَدْ عَلَّلُوهُ بِأَنَّهُ ظَنٌّ وَتَخْمِينٌ لا يُفِيدُ عِلْمًا وَلا ظَنًّا كَمَا نَقَلْنَاهُ عَنْ شَرْحِ الْبُخَارِيِّ لِلْحَافِظِ ابْنِ حَجْرٍ آنِفًا ، وَالْحِسَابُ الْمَعْرُوفُ فِي عَصْرِنَا هَذَا يُفِيدُ الْعِلْمَ الْقَطْعِيَّ... وَيُمْكِنُ لأَئِمَّةِ الْمُسْلِمِينَ وَأُمَرَائِهِمُ الَّذِينَ ثَبَتَ ذَلِكَ عِنْدَهُمْ أَنْ يُصْدِرُوا حُكْمًا بِالْعَمَلِ بِهِ فَيَصِيرُ حُجَّةً عَلَى الْجُمْهُورِ.[34]

Those who did not permit use of calculations did so because calculations [at their times] were mere guesswork and estimation serving neither factual knowledge nor proper guesswork, as we have just quoted Hafiz ibn Hajr stating that in explaining al-Bukhari. But the calculations of our times are categorical and precise in their implication. It is very much possible for Muslim leaders and rulers who confirm these categorically precise calculations to issue an edict about their application [in determining Ramadan]. This way it will become a rule/proof also for common Muslims [to follow].

Sharaf al-Quda argues that the texts permitting the use of calculations as a valid method of determining the Islamic months do not differentiate between negation and confirmation. They are generic in nature and hence good for both negation and confirmation of the months. Indeed, in his view, the hadith allowing this use proves confirmation rather than just negation.

فالنصوص الشرعية لم تفرق بين النفي و الإثبات في الأخذ بالحساب والتقدير، و بخاصة حديث (فإن غم عليكم فاقدروا له) ففي الحديث أمر بالتقدير لإثبات الشهر، و ليس لنفي الشهادة، و أما علميا فلا فرق في دقة الحساب وقطعيته بين حساب إثبات دخول الشهر، و حساب نفي دخوله. و هكذا فإن الراجح في عصرنا أن اعتماد التقدير والحساب يكون للنفي والإثبات سواء بسواء.[35]

The Islamic texts did not differentiate between confirming and negating the months by calculations. In particular, the hadith "if it is cloudy, then calculate it" commands the confirmation of the month by calculations, rather than its negation. Scientifically, it does not matter whether we use them for confirmation or negation, for they are precise and accurate anyway. Therefore, it is preferable in our time to depend equally upon calculations for confirming and negating the months.

Mustafa ʿAbd al-Basit concludes that following calculations was the original intention of the Shariʿah. Naked-eye sighting was prescribed at a time when the Ummah did not have the knowledge of precise calculations. This rule must give way to the original rule, now that the Ummah has acquired authentic knowledge of calculations. The Islamic months must be confirmed by calculation to avoid the problems connected with naked-eye sighting.[36]

The European Fiqh Council, whose president is Yusuf al-Qaradawi, includes such internationally respected jurists and scholars as Faysal Mowlawi, ʿAbd Allah ibn al-Judayʿ, and Bin Bayyah. In May 2007, it issued a fatwa that calculations for the new moon's birth are universally accepted by all scientists. Therefore, it constitutes a valid Islamic reason to confirm the lunar Islamic month the next day if the conjunction takes place and the moon sets even one second after sunset, as Mahmud Shakir said in 1939. The council established Makkah as the convention. Here is the text of its fatwa:

وأعطى المجلس فقرة خاصة للاعتناء بموضوع (الأهِلَّة)، وذلك من أجل إصدار صيغة مناسبة تدفع عن المسلمين في أوروبا العنت وتحد من الاختلاف الذي يتكرر وروده كل عام، خصوصاً في بدء شهري رمضان وشوال، وتم تناول ذلك بأبحاث خاصة في الموضوع، وهي حسب ما يلي:

1 – تعيين أوائل الشهور القمرية بين الرؤية والحساب. للدكتور محمد الهواري.

2 – ثلاث مسائل حول الهلال. للشيخ عبدالله الجديع.

3 – السبب الشرعي لوجوب صيام رمضان: هل هو دخول الشهر أم رؤية الهلال؟ للشيخ فيصل مولوي.

4 – رؤية علمية وتربوية حول رؤية الأهلة. للدكتور صلاح سلطان

قرار 4/17

إثبات دخول الشهور القمرية
استعرض المجلس مجموعة من الأبحاث بخصوص هذا الموضوع، وقرر بعد المناقشات المستفيضة ما يلي:

أن الحساب الفلكي أصبح أحد العلوم المعاصرة التي وصلت إلى درجة عالية من الدقة بكل ما يتعلق بحركة الكواكب السيارة وبخاصة حركة القمر والأرض ومعرفة مواضعها بالنسبة للقبة السماوية، وحساب مواضعها بالنسبة لبعضها البعض في كل لحظة من لحظات الزمن بصورة قطعية لا تقبل الشك.

أن لحظة اجتماع الشمس والأرض والقمر أو ما يعبر عنها بالاقتران أو الاستسرار أو المحاق لحظة كونية تحصل في لحظة واحدة، ويستطيع علم الفلك أن يحسب وقتها بدقة فائقة بصورة مسبقة قبل وقوعها لعدد من السنين، وهي تعني انتهاء الشهر المنصرم وابتداء الشهر الجديد فلكياً. والاقتران يمكن أن يحدث في أي لحظة من لحظات الليل والنهار.

يثبت دخول الشهر الجديد شرعياً إذا توافر ما يلي:

أولًا: أن يكون الاقتران قد حدث فعلًا.

ثانياً: أن يتأخر غروب القمر عن غروب الشمس ولو بلحظة واحدة مما يعني دخول الشهر الجديد، وهو قول قال به علماء معتبرون ويتوافق مع الظواهر الفلكية المعتبرة.

ثالثاً: اختيار موقع مكة المكرمة الجغرافي أساساً للشرطين المذكورين.

على البلاد الأوروبية أن تأخذ بهذه القاعدة في دخول الشهور القمرية والخروج منها وبخاصة شهرا رمضان وشوال وتحديد مواعيد هذه الشهور بصورة مسبقة، مما يساعد على تأدية المسلمين عباداتهم وما يتعلق بها من أعياد ومناسبات وتنظيم ذلك مع ارتباطاتها في المجتمع الذي تعيش فيه.

يوصي المجلس أعضاءه وأئمة المساجد وعلماء الشريعة في المجتمعات الإسلامية وغيرها بالعمل على ترسيخ ثقافة احترام ما انتهى إليه القطعي من علوم الحساب الفلكي عندما يقرر عدم إمكانية الرؤية، بسبب عدم حدوث الاقتران، أن لا يُدعى إلى ترائي الهلال، ولا يقبل ادعاء رؤيته.

سيصدر المجلس – إن شاء الله – تقويماً سنوياً يحدد بداية الشهور القمرية ونهايتها استناداً إلى هذا القرار.

The translation of this document is as follows:

The council designated a special session for the question of *hilāl* so as to issue a suitable statement that would remove the hardships faced by Muslims in Europe and limit the differences that are repeated every year, especially concerning the beginning of the months of Ramadan and Shawwal. This was achieved with research papers focusing on this question, which dealt with the following:

1. Determination of the beginning of the lunar months between sighting and calculation – Dr. Muhammad al-Hawwari.

2. Three questions concerning the *hilāl* – Shaykh ʿAbd Allah al-Judayʿ.

3. The *sharʿī* reason for the mandatory fasting of Ramadan: Is it the beginning of the month or the sighting of the crescent? – Shaykh Faysal Mowlawi.

4. A scientific and educational view of *hilāl* sighting – Dr. Salah Sultan.

Based on what was presented and discussed at this conference, the council issued the following:

Decision No. 17/4

Ascertaining the beginning of the lunar months

The council reviewed a number of research papers concerning this question and, after detailed discussions, decided the following:

1. Astronomical calculations have become one of the contemporary sciences that attained a high degree of accuracy concerning all that relates to the movements of planets, especially the movements of the moon and the earth, also their positions in relation to the celestial sphere and to one another at any time in a way that leaves no room for doubt.

2. That the moment of the lining up of the sun, moon, and earth – or what is referred to as the conjunction – is a cosmic point that occurs in one moment. It is possible for the science of astronomy to calculate that moment with extreme accuracy several years before it occurs. It [the conjunction] means the end of the previous astronomical month and the beginning of the following month. Conjunction may occur at any moment, day or night.

The beginning of the *sharʿī* new month is ascertained if the following conditions are satisfied:

(a) **First:** The conjunction must have actually occurred.

(b) **Second:** The moonset follows sunset even by one moment, which marks the beginning of the month. This view was expressed by credible scholars and is consistent with the credible astronomical phenomena.

(c) **Third:** The choice of the geographic position of Makkah as the basis of the two above conditions.

4. [Muslims in] the European countries should adopt this rule to determine the beginning and end of lunar months, especially the months of Ramadan and Shawwal, and to determine these dates in advance so as to help Muslims to perform their acts of worship and their related festivals and occasions, also to regulate their connection with the communities in which they live.

5. The council exhorts it members, imams of mosques, and Shariʿah scholars in Muslim societies and other societies to inculcate the culture of respect of what has been concluded by the definitive aspects of astro-

nomical calculation when it is determined that the sighting [of the new *hilāl*] is impossible, since the conjunction had not yet occurred. No call for attempting to sight the *hilāl* should be made and any claim of sighting should not be accepted.[37]

CONCLUSIONS

1. The assertion that there exists a juristic consensus banning the use of precise astronomical calculations in confirming or negating Ramadan is unfounded. However, the majority of *jumhūr* adopted that opinion due to the uncertain nature of calculations during their time and also because of the possible negative ramifications of that in areas related to faith and *ᶜaqīdah*.

2. A group of well-known authorities in three Sunni legal schools, with the exception of the Hanbalis, has from early times argued in favor of accepting calculations in part or in totality.

3. Modern science has attained such a level of authenticity in calculation that achieving certainty about the birth, presence, and absence of the moon on the horizon and so on is not difficult at all. This scientific method is definitely more trustworthy than naked-eye sighting.

4. The number of scholars inclining toward the partial or total acceptance of calculation to confirm Ramadan is increasing daily, mostly owing to the certainty and ease of this method. This method also results in many communal, financial, and social benefits, the preservation of which is an integral part of the Shariᶜah's objectives.

5. Some very conservative modern Salafi/Hanbali scholars, such as Mahmud Shakir, have also accepted this point of view. Indeed, scholars like Mahmud Shakir and Rashid Rida advocate calculation as the only authentic and lawful method currently available to Muslims vis-à-vis following the Sunnah's true essence regarding the fast of Rama-dan. Mahmud Shakir has done this since 1939.

6. The new moon is just a sign of sacred timings. The moon has a beginning and a clear end in its orbit around the earth. The beginning is its birth and is the most certain point that can be determined, even months and years ahead of its occurrence, with the help of calculations. Thus there is nothing wrong in accepting the new moon's birth as the con-

vention to start the new month. In fact, this is the only authentic and certain convention to determine the new month. The criteria of visibility are not agreed upon even by Muslim astronomers and scholars. Once it has been established that certainty about Ramadan and the fasting during that month, and not the actual sighting of its moon, is the intended cause of fasting as well as the objective of the Shari^cah, then disputing the question of visibility and non-visibility will become a fruitless endeavor. We should take the birth as the accepted norm and announce the Islamic calendar long ahead of time, as has been done by the fiqh councils in North America and Europe.

7. Greenwich Mean Time is an arbitrary convention accepted by the international community to facilitate times and dates. It has no Islamic value whatsoever. On the other hand Makkah, being the sanctuary of all Muslims, enjoys greater significance than GMT. Therefore, Muslims should take Makkah as the Islamic convention to determine the Islamic lunar months. The new month will start when the new moon is born before sunset in Makkah and stays in the horizon after sunset, even this event only lasts for a short time. The whole Muslim world would then have the beginning of the new month within twenty-four hours of the new moon's birth in Makkah.

Given all of the above, accepting calculations to both confirm and negate Ramadan conforms to the true essence of the Sunnah and does not constitute any deviation from the spirit of the Shari^cah. In contrast, it is perhaps the only available method that, if applied in spirit, can realize the Islamic objectives of authenticity, certainty, and unity. وَٱللَّهُ أَعْلَمُ.

ENDNOTES

PREFACE

1. Abu ʿAbd Allah Muhammad ibn Ismaʿil al-Bukhari, *Ṣaḥīḥ al-Bukhārī*, ed. Muhammad Zahir ibn Nasir al-Nasir (Beirut: Dar Tawqa al-Najat, 1422 AH) vol.6, p.481.
2. Ibid., vol.6, p.479.
3. Ibid., vol.6, p.487.
4. Ibid., vol.1, p.423.

INTRODUCTION

1. See Mokhtar Maghraoui, "An Islamic Legal Analysis of the Astronomical Determination of the Beginning of Ramadan," August 2007, www.zawiyah.net/artop; also Hamza Yusuf, *Caesarean Moon Births: Calculations, Moon Sighting, and the Prophetic Way* (Berkeley: Zaytuna Institute, 2007).
2. Hamza Yusuf, in particular, emphasized this point. See *Caesarean Moon Births,* pp.31-32.
3. Ismail Poonawala, "Ramadan," http://encarta.msn.com/encyclopedia_761-5579768/ramadan.html.
4. Ibid.
5. Ibid.
6. Encarta, "Copernican System," http://encarta.msn.com/encyclopedia_761-557138/copernican_system.html.
7. Ibid., "Astronomy," p.195.

CHAPTER 1

1. Al-Bukhari, *Ṣaḥīḥ al-Bukhārī*, vol.3, p.499.
2. Abu al-Hasan Muslim ibn al-Hajjaj al-Qushayri, *Ṣaḥīḥ Muslim*, ed. Fu'ad ʿAbd al-Baqi (Beirut: Dar Ihya' al-Turath al-ʿArabi, n.d.), vol.7, p.227.
3. Zayn al-Din ʿAbd al-Rahman ibn Ahmad ibn Rajab al-Hanbali, *Fatḥ al-Bārī* (Dammam: Dar Ibn al-Jawzi, 1422 AH), vol.3, p.142.
4. Muhammad Amin ibn ʿUmar ibn ʿAbidin, *Radd al-Mukhtār ʿalā al-Durr al-Mukhtār* (Beirut: Dar al-Fikr, 1421 AH) vol.7, p.336.
5. Muhammad Rashid ibn ʿAli Rida, *Tafsīr al-Qur'ān al-Ḥakīm* (viz., *Tafsīr al-Manār*) (Cairo: al-Hayat al Misriyyah al-ʿAmmah li al-Kitab, 1990), vol.2, pp.149-50.

6. Ibid., vol.2, p.151.
7. Ibn Rajab, *Fatḥ al-Bārī*, vol.3, p.142.
8. Ibid., vol.3, p.142.
9. Ibid.
10. *Al-Mawsūʿah al-Fiqhīyah* (Kuwait: Ministry of Islamic Affiars and Endowments, 1967), vol.23, p.142 (various printings).
11. ʿAbd Allah ibn Ahmad ibn Muhammad ibn Qudamah, *Al-Mughnī* (Beirut: Dar al-Fikr, 1405 AH), vol.44, p.324.
12. Ahmad ibn Idris al-Qarrafi, *Anwār al-Burūq fī Anwāʿ al-Furūq* (Beirut: Dar al-Kutub al-ʿIlmiyyah, 1424 AH), vol.1, p.19.
13. ʿUthman ibn ʿAli al-Zayla'i, *Tabyīn al-Ḥaqā'iq Sharḥ Kanz al-Daqā'iq* (Beirut: Dar al-Kitab al-Islami, n.d.), vol.1, p.316; see also Ibn ʿAbidin, *Radd al-Mukhtār*, vol.2, p.619.
14. Al-Zayla'i, *Tabyīn*, vol.1, p.321.
15. Abu Muhammad ʿAbd Allah ibn Abi Zayd al-Qayrawani, *Al-Thamar al-Dāni', Risālah Abī Zayd al-Qayrawānī* (Beirut: Dar al-Fikr, n.d.), vol.1, p.240.
16. Muhammad ibn ʿAli al-Shawkani, *Nayl al-Awṭār* (Beirut: Dar al-Jil, 1973), vol.4, p.269.
17. Al-Qarrafi, *Anwār al-Burūq*, vol.4, p.142 .
18. Ibid., vol.1, p.17; *Al-Mawsūʿah al-Fiqhīyah*, vol.22, p.36.
19. *Al-Mawsūʿah al-Fiqhīyah*, vol.22, p.36.
20. Al-Qarrafi, *Anwār al-Burūq*, vol.4, p.143.
21. Muhyiddin Abu Zakariyyah Yahya ibn Sharaf al-Nawawi, *Al-Majmūʿah Sharḥ al-Muhadhdhab* (Cairo: Matbaʿah al-Muniriyyah, n.d.), p.6.
22. Ibid., vol.6, p.282.
23. Al-Zayla'i, *Tabyīn*, vol.1, p.316.
24. Ibn ʿAbidin, *Radd al-Mukhtār*, vol.2, p.393.
25. ʿAbd Allah ibn Ahmad al-Nasafi, *Kanz al-Daqā'iq* (Beirut: Dar al-Nashr, n.d.), p.17; al-Zayla'i, *Tabyīn*, vol.1, p.321.
26. Al-Zayla'i, *Tabyīn*, vol.4, p.78.
27. *Ṣaḥīḥ Muslim*, vol.5, p.367.
28. Muhammad Shams al-Haqq al-Azimabadi, *ʿAwn al-Maʿbud Sharḥ Abī Dāwūd* (Beirut: Dar al-Kutub al-ʿIlmiyyah, 1415 AH), vol.6, p.453.
29. Qadi Abu Bakr Muhammad ibn ʿAbd Allah ibn al-ʿArabi, *Aḥkām al-Qur'ān*, edited by ʿAli Muhammad al-Bayjawi', 1st edn. (Beirut: Dar Ihya' al-Turath al-ʿArabi, n.d.), vol.1, 157.
30. Jamal al-Din ʿAbd Allah ibn Yusuf al-Zayla'i, *Naṣb al-Rāyah fī Takhrīj Aḥādīth al-Hidāyah* (Cairo: Dar al-Hadith, n.d.), vol.3, p.40.
31. Ibn Qudamah, *Al-Mughnī*, vol.4, p.324.
32. Al-Shawkani, *Nayl al-Awṭār*, vol.4, p.268.
33. Ibn Qudamah, *Al-Mughnī*, vol.4, p.324.

34. Rowan A. Greer, trans., *Origen: The Classics of Western Spirituality Series* (New York: Paulist Press, 1979), p.180.
35. Henry Chadwick, *Early Christian Thought and the Classical Tradition* (New York: Oxford University Press, 1966), p.25ff.
36. Greer, *Origen*, p.180.
37. John Ferguson, *Clement of Alexandria* (New York: Twayne Publishers, 1974), p.73ff.
38. Raymond E. Brown defines *sensus literalis* as "the sense which the human author directly intended and which his words convey." See Joseph Fitzmeyer Brown and Jerome Murphy, eds., "The Literal Sense of Scripture," *The Jerome Biblical Commentary* (Englewood Cliffs, NJ: Prentice-Hall, 1968), p.606.
39. Greer, *Origen*, p.180.
40. Brown, "The Literal Sense," p.607.
41. G. W. Butterworth, trans. *Origen on First Principles* (London: SPCK, 1936), vol.4, p.297.
42. Charles Bigg, *The Christian Platonists of Alexandria: The 1886 Bampton Lectures* (Oxford: Clarendon Press, 1968), p.185.
43. Harry A. Wolfson, *The Philosophy of the Church Fathers*, 3rd rev. edn. (Cambridge, MA: Harvard University Press, 1970), pp.57-62.
44. Bigg, *Christian Platonists*, p.185.
45. Emil Brunner, *The Christian Doctrine of God*, trans. Olive Wyon (Philadelphia: Westminster, 1974), vol.1, p.108.
46. John Bright, *The Authority of the Old Testament* (New York: Abingdon Press, 1967), pp.80-81.
47. Ibid., p.81.
48. Ibid., p.82ff.
49. Zulfiqar A. Shah, "A Study of Anthropomorphism and Transcendence in the Bible and Qur'an: Scripture and God in the Judeo Christian and Islamic Tradition," chapter 2. Submitted to the University of Wales (Lampeter) in 1997.
50. Ibn Kathir, *Al-Bidāyah wa al-Nihāyah* (Cairo: Matbaʿah al-Saʿadah, 1965), vol.9, p.350.
51. Abu al-Hasan al-Ashʿari, *Maqālāt al-Islāmīyīn wa Ikhtilāf al-Muṣallīn*, M. ʿAbd al-Hamid, ed. (Beirut: al-Hikmah, 1994), p.161.
52. Ibrahim Madkur, *Fī al-Falsafah al-Islāmīyah* (Cairo: Dar al-Maʿarif, 1976), p.2.
53. M. S. Seale, *Muslim Theology, A Study of Origins with Reference to the Church Fathers* (London: Lugac and Co., 1964), p.4.
54. Hamilton A. R. Gibb, *Mohammedanism* (London: Oxford Univeristy Press, 1972), p.42.
55. See Jar Allah Zahdi Hasan, *Al-Muʿtazilah* (Cairo: al-Mu'assasat al-ʿArabiyyah li al-Dirasat wa al-Nashr, 1947), pp.33, 256.
56. Andrew Rippin, *Muslims: Their Religious Beliefs and Practices* (New York: Routledge, 1990), vol.1, p.6.

57. Ibid.
58. Frederick M. Denny, *An Introduction to Islam*, 2nd edn. (New York: Prentice Hall: 1994), p.18.
59. See Qadi ʿAbd al-Jabbar's emphasis upon the issue in his *Al-Muḥīt bi al-Taklīf* (Cairo: al-Mu'assasat al-Masriyyah, 1965), vol.1, p.1099.
60. Ian R. Netton, *Allah Transcendent* (London and New York: Routledge, 1989), p.4.
61. See a detailed discussion of these interpretations in J. M. S. Baljon, "Qur'anic Anthropomorphism," *Islamic Studies* (Islamabad: Summer 1998), no. 27; see also al-Ashʿari, *Al-Ibānah*, pp.36-45; Qadi A. Jabbar, *Sharḥ Uṣūl al-Khamsah*, ed. ʿAbd al-Karim ʿUthman (Cairo: Maktabah Wahbah, 1965), p.227ff.
62. W. Montgomery Watt, *The Formative Period of Islamic Thought* (Edinburgh: Edinburgh University Press, 1973), p.248.
63. Georges Anawati, "Attributes of God," in *Encyclopedia of Religion*, 2nd edn., ed. Lindsay Jones, (MacMillian Reference Books: 2005), p.514.
64. Watt, *Formative Period*, p.24.
65. D. B. MacDonald, *Muslim Theology, Jurisprudence and Constitutional Theory* (Beirut: Khayats, 1965), p.14.
66. Fazlur Rahman, *Islam* (Chicago and London: University of Chicago Press, 1979), p.8.
67. Netton, *Allah Transcendent*, p.4.
68. M. Watt, "Early Discussions about the Qur'an," *Muslim World*, XL (January 1950), p.3.
69. Hamilton A. R. Gibb, *Modern Trends in Islam* (New York: Octagon Books, 1972), pp.19-2.
70. Abu Nasr al-Farabi, *Mabādī' Ārā' Ahl al-Madīnah al-Fāḍilah* (Leiden: E. J. Brill, 1890), pp.17-18; Richard Walzer has translated it into English. See *Al-Farabi on the Perfect State* (Oxford: Clarendon, 1985). See more details in al-Farabi, *Al-Thamarāt al-Marḍīyah* (Leiden: E. J. Brill, 1895), p.57ff.
71. See Arthur J. Arberry, *Avicenna on Theology* (Westport, CT: Hyperion Press, 1979), pp.25ff.
72. Abu al-Walid Muhammad ibn Ahmad ibn Muhammad ibn Rushd, *Faṣl al-Maqāl wa-Taqrīb ma bayn al-Sharīʿah wa al-Ḥikmah min al-Ittiṣāl*, ed. G. F. Hourani (Leiden: E. J. Brill, 1959); see also ʿAbd al-Halim Mahmud, *Al-Tafkīr al-Falsafī fī al-Islām* (Cairo: Dar al-Maʿarif, 1984).
73. See details of their philosophical positions in Seyyed Hossein Nasr and Oliver Leaman, eds., *History of Islamic Philosophy* (New York: Routledge, 1996), vol.1, pp.178-97, 231-51, and 330-34.
74. See Azim Nanji, "Ismaʿili Philosophy," in ibid., vol.1, pp.144-45.
75. See Hamid al-Din al-Kirmani, *Rāḥat al-ʿAql* (Cairo: Dar al-Fikr al-ʿArabi, 1952), p.46; also Arif Tamir, *Khams Rasā'il Ismāʿīlīyah* (Damascus: Dar al-Insaf, 1956); and *Ḥaqīqat Ikhwān al-Ṣafā' wa Khalān al-Wafā'* (Beirut:

Catholic Press, 1957); Adil al-Awa, *Muntakhabāt Ismāʿīlīyah* (Damascus: Matbʿah al-Jamiʿah al-Suriyyah, 1957).
76. Netton, *Allah Transcendent*, p.203.
77. Abu Hanifah al-Nuʿman, *Al-Fiqh al-Akbar*, ed. Mulla ʿAli al-Qari (Cairo: Al-Babi, 1955). For the English translation and commentary, see A. J. Wensick, *The Muslim Creed* (London: Frank Cass & Co. Ltd., 1965), pp.102-247. Some scholars like Watt attribute this book to some later (between 900-950 AC) Hanafi sources.
78. See Ahmad ibn Hanbal, *Al-Radd ʿalā al-Zandiqah wa al-Jaḥmīyah* (Cairo: Maktabah al-Salafiyyah, 1393 AC).
79. Majid Fakhry, *A History of Islamic Philosophy* (New York: Columbia University Press, 1970), p.3; also see Muhammad ibn ʿAli al-Shawkani, *Al-Tuhaf fī Madhāhib al-Salaf* (Cairo: Matbaʿah al-Imam, n.d.).
80. E. Sell, "God" (Muslim) in *Encyclopeadia of Religion and Ethics*, ed. James Hastings (New York: T. & T Clark Publishers, 1995) p.30.
81. ʿAbd al-Karim al-Shahrastani, *Al-Milal wa al-Niḥal* (London: Cureton's edition, 1842), p.76.
82. W. M. Watt, *Islamic Philosophy and Theology* (Edinburgh; Edinburgh University Press, 1967), p.8.
83. Ibid., p.8.
84. Karen Armstrong, *A History of God* (New York: A. A. Knopf, 1994), p.16.
85. A. K. Kazi and J. G. Flynn, "Muslim Sects and Divisions," *Kitāb al-Milal wa al-Niḥal by Shahrastānī*, trans. A. K. Kazi and J. G. Flynn (London: Kegan Paul International, 1984), p.8.
86. See details in Ibn Taymiyyah, *Majmūʿ Fatāwā*, ed. ʿAbd al-Rahman ibn Muhammad ibn Qasim (Ribat: Maktabah al-Maʿarif, n.d.), vol.5, p.323ff.
87. Nasr and Leaman, eds., *Islamic Philisophy*, pp.113-15.
88. See Ibn Taymiyyah, *Minhāj al-Sunnah fī Naqdī Kalām al-Shīʿah wa al-Qadarīyah* (Beirut: Dar al-Fikr, n.d.); *Naqd al-Manṭaq*, 1st edn. (Cairo: Matbaʿah al-Sunnah al-Muhammadiyyah, 1951); *Al-Risālah al-Tadummurīyah* (Beirut: al-Maktab al-Islami, n.d.); *Mawafaqah Ṣarīḥ al-Maʿqūl li Ṣaḥīḥ al-Manqūl* (Cairo: 1321 AH); *Al-Rasā'il wa al-Masā'il* (Beirut: al-Maktab al-Islami, n.d.); and *Kitāb al-Nubuwwat* (Cairo: al-Maktabah al-Salafiyyah, 1382 AH).
89. M. Nasir al-Din al-Albani, *Mukhtaṣar al-ʿUlūw* (Beirut: al-Maktab al-Islami, 1412 AH) 2007, vol.1, p.36.
90. Al-Qarrafi, *Anwār al-Burūq*, vol.2, p.229.

CHAPTER 2

1. Abu al-Thana' Muhammad Shihab al-Din Al-Alusi, *Rūḥ al-Maʿānī fī Tafsīr al-Qur'ān al-ʿAẓīm wa al-Sabʿ al-Mathāni* (Beirut: Dar Ihya' al-Turath al-ʿArabi, n.d.), vol.2, p.129.

2. Ahmad ibn Faris Abu al-Hussayn al-Qazwini al-Razi, *Maqāyīs al-Lughah*, ed. ʿAbd al-Salam Muhammad Harun (Damascus: Ittihad al-Kuttab al-ʿArab, 2000).
3. Fakhr al-Din Abu ʿAbd Allah Muhammad al-Razi, *Tafsīr al-Rāzī* (Beirut: Dar al-Kutub al-ʿIlmiyyah, 1978), vol.4, p.142.
4. Jalal al-Din al-Suyuti, *Tafsīr al-Jalālayn* (Beirut: Dar Ihya' al-Turath al-ʿArabi, n.d.).
5. Abu ʿAbd Allah al-Qurtubi, *Tafsīr al-Qurṭubī* (Beirut: Dar al-Kutub al-ʿIlmiyyah, n.d.), vol.2, p.290.
6. ʿImad al-Din Hafiz ibn Kathir, *Tafsīr al-Qur'ān al-ʿAẓīm* (Beirut: Dar Ihya' al-Turath, n.d.), vol.1, p.360.
7. Al-Suyuti, *Tafsīr Jalālayn*, p.37.
8. ʿAbd Allah ibn Ahmad al-Nasafi, *Tafsīr al-Nasafī* (Beirut: Dar Ihya' al-Turath, n.d.).
9. Al-Razi, *Tafsīr*, vol.2, p.250.
10. Al-Alusi, *Tafsīr*, vol. 2, p.250.
11. Al-Alusi, *Rūḥ al-Maʿānī*, vol.2, p.250.
12. Al-Alusi, *Tafsīr*, vol.2, p.250.
13. Al-Razi, *Tafsīr*, vol.4, p.142.
14. Ibid., vol.3, p.103.
15. Zaheer Uddin, "A Refutation to ISNA/Fiqh Council's Decision to Disregard the Qur'an and the Sunnah and to Follow Astronomical Calculations for Beginning an Islamic Month," www.hilalsighting.org. Issued on Rajab 29, 1427; August 24, 2006.
16. Ibn ʿAbbas' statement could have multiple interpretations. Al-Tabari reports in his *Tafsīr*, vol.3, p.144:
17. Quoted from Hamza Yusuf, *Caesarean Moon Births*, p.28.
18. Abu Bakr al-Jassas, *Aḥkām al-Qur'ān* (Beirut: Dar al-Fikr, n.d.), vol.1, p.117.
19. Yusuf, *Caesarean Moon Births*, vol.1, p.15.
20. Ibid., vol.1, p.15.
21. Abu Bakr Muhammad ibn ʿAbd Allah ibn al-ʿArabi, *Aḥkām al-Qur'ān* (Beirut: Dar al-Kutub al-ʿIlmiyyah, n.d.), vol.1, p.152.
22. Yusuf, *Caesarean Moon Births*, pp.23-24. Note: Italics are added.
23. Faysal Mowlawi, "Al-Sabab al-Sharʿī li Wujūb Ṣiyām Ramaḍān: Hal Huwa Dukhūl al-Shahr am Ru'yat al-Hilāl?" an unfinished paper presented to the European Fiqh Council in May of 2007.
24. Al-Jassas, *Aḥkām al-Qur'ān*, vol.1, p.279.
25. Ibid., vol.1, p.456.
26. Ibid., vol.2, p.47.
27. Ibid., vol.1, p.456.
28. Ibid., vol.1, p.496.
29. Mowlawi, "*Al-Sabab al-Sharʿī*," p.24.
30. Ibid., p.24.

31. Al-Qarrafi, *Anwār al-Burūq*, vol.4, pp.139-40.
32. Al-Qurtubi, *Tafsīr*, vol.1, p.1.
33. Ibid., vol.19, p.22.
34. See Ibn Manzur, *Lisān al-ʿArab* (Beirut: Dar Ihya' al-Turath al-Islami, n.d.), vol.11, p.701.
35. Muhammad ibn Yaʿqub al-Fayruzabadi, *Qāmūs al-Muḥīṭ* (Beirut, Mu'assasat al-Risalah, n.d.), vol.3, p.183.
36. Ibn Manzur, *Lisān al-ʿArab*, vol.11, p.701.
37. Ibid.
38. Ibid.
39. Ibid.
40. Abu Hayyan al-Andalusi, *Tafsīr al-Baḥr al-Muḥīṭ* (Beirut: Dar al-Fikr, n.d.), Ayah 189, vol.2, p.215.
41. Al-Jawhari, *Al-Ṣiḥāḥ fī al-Lughah* (Beirut: Dar al-ʿIlm li al-Malayin, 1990), vol.2, p.254.
42. Abu Mansur Muhammad ibn Ahmad al-Harawi al-Azhari, *Tahdhīb al-Lughah* (Beirut: Dar Ihya' al-Turath al-ʿArabi, n.d.), vol.2, p.225; Ibn Manzur, *Lisān al-ʿArab,* vol.11, p.701.
43. Abu al-Qasim Mahmud ibn ʿUmar Jar Allah al-Zamakhshari, *Asās al-Balāghah* (Beirut: Dar Ihya' al-Turath al-ʿArabi, n.d.), vol.2, p.4.
44. Al-Alusi, *Rūḥ al-Maʿānī*, vol.2, p.142.
45. Ahmad ibn Muhammad ibn ʿAli al-Fayyumi, *Al-Miṣbāḥ al-Munīr fī Gharīb al-Sharḥ al-Kabīr* (Cairo: Matbaʿah al-Taqaddum al-ʿIlmiyyah, 1323 AH), vol.10, p.169.
46. Yusuf, *Caesarean Moon Births*, p.18.
47. Ibid., p.19.
48. Ibid. Italics are added.
49. Ibid.
50. Al-Khalil ibn Ahamd al-Farahidi, *Al-ʿAyn* (Beirut: Dar al-Kutub al-ʿIlmiyyah, n.d.) vol.1, p.245.
51. Ibid., vol.1, p.246.
52. Ibid.
53. Yusuf, *Caesarean Moon Births*, p.18.
54. Ibn Taymiyyah, *Majmūʿah Fatāwā Ibn Taymīyah* (al-Mansurah: Dar al-Wafa', 2005), vol.6, p.69.
55. Abu Jaʿfar Muhammad ibn Jarir al-Tabari, *Jāmiʿ al-Bayān fī Ta'wīl al-Qur'ān, Tafsīr al-Ṭabarī* (Beirut: Mu'assasat al-Risalah, 2000), vol.2, p.224.
56. Abu al-Fadal Shihab al-Din Ahmad ibn ʿAli ibn Hajar, *Fatḥ al-Bārī* (Beirut: Dar al-Fikr, n.d.), vol.5, p.193.
57. Shaykh Maghraoui states: "*Hilāl* traditionally referred to a crescent that is at least one or nights old…"; Here the word "traditionally" is important to notice. See Maghraoui, *An Islamic Legal Analysis*, p.7.

58. See Taqi al-Din ibn al-Najjar, *Sharḥ al-Kawkab al-Munīr* (Riyadh: Maktabat al-ʿAbaykan, 1997), vol.4, p.448-49.
59. Abu Muhammad ʿAbd al-Haqq ibn Ghalib ibn ʿAtiyyah, *Al-Muḥarrar al-Wajīz* (Beirut: Dar al-Kutub al-ʿIlmiyyah, 1993), vol.3, p.139.
60. Ibn al-Najjar, *Sharḥ al-Kawkab*, vol.3, p.41.
61. Ali Hayder Khawaja Afindi, *Durar al-Ḥikām Sharḥ Majallah al-Aḥkām*, tr. Fahmi al-Hussaini (Beirut: Dar al-Kutub al-ʿIlmiyyah, n.d.), vol.1, p.70.
62. Ibid., vol.1, p.70.
63. For details, see al-Tabari, *Tafsīr;* al-Alusi, *Tafsīr al-Alusī.*
64. Al-Razi, *Al-Tafsīr al-Kabīr* (*Tafsīr al-Rāzī*) (Beirut: Dar al-Fikr, 1978), vol.8, p.20.
65. Ibid., vol.8, p.21.
66. Tracy R. Rich, "The Jewish Calendar: A Closer Look," www.JewFAQ.org/calendar.htm.
67. Al-Razi, *Tafsīr*, vol.8, p.21.
68. Yusuf, *Caesarean Moon Births*, p.22.
69. Ibid.
70. Abu Dawud, *Sunan Abū Dāwūd* (Beirut: Dar al-Fikr, n.d.), vol.13, p.286.
71. Abu ʿAbd Allah Ahmad ibn Muhammad ibn Hanbal, *Musnad Aḥmad* (Cairo: Mu'assast Qurtubah, n.d.), vol.3, p.332.
72. Abu Dawud, *Sunan Abū Dāwūd*, vol.13, p.287.
73. Ibid., vol.13, p.436.
74. Ibn al-ʿArabi, *Aḥkām al-Qur'ān* (Beirut: Dar al-Fikr, 1978), vol.1, p.159.
75. See more details in Mulla ʿAli Qari, *Mirqāt al-Mafātīḥ* (Beirut: Dar al-Fikr, n.d.), vol.5, p.282.

CHAPTER 3

1. Maghraoui, *An Islamic Legal Analysis,* p.8.
2. Rashid Rida, *Tafsīr al-Manār* (Beirut: Dar al-Kutub al-ʿIlmiyyah, 1999), vol.2, p.149-50.
3. Ibid., vol.2, p.186.
4. Al-Bukhari, *Ṣaḥīḥ al-Bukhārī*, vol.6, p.481.
5. Muslim, *Saḥīḥ Muslim* (Beirut: Dar Ihya' Turath al-ʿArabi, n.d.), vol.5, p.355.
6. Al-Bukhari, *Saḥīḥ,* vol.6, p.478.
7. *Musnad Aḥmad*, vol.11, p.78.
8. Yusuf, *Caesarean Moon Births*, pp.3, 4; italics are added.
9. Ibid.; italics added for emphasis.
10. Abu ʿIsa Muhammad al-Tirmidhi, *Sunan al-Tirmidhī* (Beirut: Dar Ihya' Turath al-ʿArabi, n.d.), vol.3, p.113.
11. Abu Muhammad ʿAbd Allah ibn ʿAbd al-Rahman al-Darimi (Beirut: Dar al-Kitab al-ʿArabi, 1407 AH), vol.5, p.166.

12. Abu Dawud, *Sunan*, vol.6, p.263.
13. Al-Bukhari, *Saḥīḥ*, vol.6, p.478.
14. *Musnad Aḥmad*, vol.11, p.478.
15. Ibn Daqiq al-ʿId, *Iḥkām al-Aḥkām Sharḥ ʿUmdat al-Aḥkām* (Cairo: Matbaʿah al-Sunnah al-Muhammadiyyah, n.d.), vol.2, p.8.
16. Yusuf, *Caesarean Moon Births*, p.44.
17. Ibid.
18. Ibn Daqiq al-ʿId, *Iḥkām al-Aḥkām*, vol.2, p.8.
19. Ibid.
20. Masʿud ibn ʿUmar al-Taftazani, *Sharḥ al-Talwīḥ ʿalā al-Tawḍīḥ* (Cairo: Maktabah Sabih, n.d.), vol.1, p.401.
21. Mustafa al-Zarqa, *Fatāwā Muṣṭafā al-Zarqā*, ed. Majd Ahmad Makki, 2nd edn. (Damascus: Dar al-Qalam, 1422 AH/2001 AC), pp.163-64.
22. Maghraoui, *An Islamic Legal Analysis*, p.7; italics are added.
23. *Ṣaḥīḥ al-Bukhārī*, vol.7, p.57.
24. Muslim, *Ṣaḥīḥ*, vol.5, p.395.
25. Abdur Rahman Ijaz, "Axing of Methodology," www.hilalcommittee.com/Axing_of_Methodology_rev1_2.pdf, p.4; some words are inserted to complete the sentences and clarify the meanings.
26. Al-Tabari, *Tafsīr al-Ṭabarī*, vol.10, p.21.
27. Ibid.
28. Bin Bayyah, www.Binbayyah.net, section on "Fatāwā"; this is not to say that Shaykh Bin Bayyah endorses calculations for confirmation of Ramadan. In this fatwa he seems to be maintaining use of calculations in proving or reaffirming the sighting.
29. Zayn al-Din ibn Ibrahim Ibn Nujaym, *Al-Baḥr al-Rāqī Sharḥ Kanz al-Daqā'iq* (Beirut: Dar al-Kitab al-Islami, n.d.), vol.2, p.284.
30. Taqi al-Din al-Subki, *Fatāwā al-Subkī* (Cairo: Dar al-Maʿarif, n.d.), vol.1, p.211.
31. Yusuf al-Qaradawi, *Fatāwā Muʿāṣirah* (Damascus: Dar al-Qalam, 1996), vol.2, pp.212-17.
32. *Al-Mawsūʿah al-Fiqhīyah*, vol.14, p.53.
33. Ahmad ibn Muhammad al-Hamwi, *Ghamz ʿAwn al-Baṣā'ir* (Beirut: Dar al-Kutub al-ʿIlmiyyah, n.d.), vol.2, p.66.
34. Al-Zarqa, *Fatāwā*, pp.157-59.
35. Yusuf, *Caesarean Moon Births*, pp.31, 32.
36. Ibid., p.32.
37. Ibid., p.45.
38. Encarta, "Solar System," http://encarta.msn.com/encyclopedia_761557663/solar_system.html.
39. Yusuf, *Caesarean Moon Births*, p.45.
40. Al-Qarrafi, *Anwār al-Burūq*, vol.4, pp.144-45.
41. Ibid., vol.4, pp.140-41.

42. Al-Shatibi, *Al-Muwāfaqāt*, vol.1, p.54.
43. ʿAbd al-Karim Zaydan, *Al-Wajīz fī Uṣūl al-Fiqh* (Cairo: Dar al-Tawziʿ wa al-Nashr al-Islamiyyah, 1414 AH/1993 AC), p.55.
44. Zaydan, 26; also see Faysal Mowlawi, *Al-Sabab al-Sharʿī*, p.8.
45. Zaydan, *Al-Wajīz*, pp.26-27.
46. Al-Nawawi, *Al-Majmūʿah Sharḥ al-Muhadhdhab* (Cairo: Matbaʿat al-Muniriyyah, n.d.), vol.6, p.270.
47. Cited from Yusuf, *Caesarean Moon Births*, p.24.
48. Al-Razi, *Tafsīr al-Rāzī* (Beirut: Dar al-Fikr, 1978), vol.3, p.103.
49. Mowlawi, *Al-Sabab al-Sharʿī*, p.18.
50. See detail about *sabab* and *ʿillah* in ibid., p.8ff.
51. Muhammad Abi Zahrah, *ʿIlm Uṣūl al-Fiqh* (Cairo: Dar al-Fikr al-ʿArabi, 1998), p.56.
52. ʿAbd Allah ibn Yusuf al-Judayʿ, *Taysir ʿIlm Uṣūl al-Fiqh* (Beirut: Mu'assasat al-Rayyan, 2005), p.54.
53. ʿAbd al-Karim Zaydan, *Al-Wajīz*, p.55.

CHAPTER 4

1. See W. B. Smith, *The Birth of the Gospel: A Study of the Origin and Purport of the Primitive Allegory of the Jews* (New York: Philosophical Library, 1957).
2. See Stephen Neil, *The Interpretation of the New Testament 1861-1961* (New York: Oxford University Press, 1966).
3. V. Paul Furnish, *Jesus According to Paul* (Cambridge: Cambridge University Press, 1993), p.40; also see H. Anderson, *Jesus and Christian Origins* (New York: Oxford University Press, 1964); E. P. Sanders, *The Historical Figure of Jesus* (England: Penguin, 1993).
4. John Hick, *The Metaphor of God Incarnate* (London: SCM Press, 1993), p.31.
5. Readers interested in the historical development of "Jesus Doctrine" are reffered to my Ph.D. dissertation, "A Study of Anthropomorphism and Transcendence in the Bible and Qur'an: Scripture and God in the Judeo Christian and Islamic Tradition." Submitted to the University of Wales (Lampeter) in 1997, also available at www.hilal-discourse.net.
6. Ibid., chapter 4 for details.
7. Abdulaziz Sachedina, "Warfare: The Use and Abuse of Jihad in the Muslim World," www.uga.edu/islam/jihad; see also www.islamiclearning.org.
8. Khaled Abou El Fadl, "Speaking, Killing, and Loving in God's Name," *The Hedgehog Review* (spring 2004); www.scholarofthehouse.org/skiandloingo.html; Khaled Abou El Fadl, "The Rules of Killing at War: An Inquiry into Classical Sources," *The Muslim World*, no. 89 (1999).
9. Ibn Hajar, *Fatḥ al-Bārī*, vol.6, p.147.
10. Ibn Qudamah, *Al-Mughnī*, vol.3, p.7.

11. *Al-Mawsūʿah al-Fiqhīyah*, vol.22, p.32.
12. Al-Subki, *Fatāwā*, vol.1, p.217.
13. Al-Qaradawi, *Fatāwā Muʿāṣirah*, vol.2, p.222.
14. Ahmad Muhammad Shakir, *Awā'il al-Shuhūr al-ʿArabīyah* (Cairo: Maktabah Ibn Taymiyyah, 1987).
15. Abu Bakr ibn ʿAli al-Razi al-Jassass, *Aḥkām al-Qur'ān* (Beirut: Dar al-Fikr, n.d.), vol.1, p.279.
16. Ibid., vol.1, p.280.
17. Ibid.
18. Badr al-Din al-ʿAyni, *ʿUmdat al-Qārī* (Beirut: Dar al-Fikr, n.d.), vol.10, p.265.
19. Al-Jassas, *Aḥkām*, vol.1, p.117.
20. Al-Hamwi, *Ghamz ʿAwn al-Baṣā'ir*, vol.2, p.66.
21. Muhammad ibn ʿAbd Allah al-Kharshi, *Sharḥ Mukhtaṣar Khalīl li al-Kharshī* (Beirut: Dar al-Fikr, n.d.), vol.2, p.237.
22. Muhammad ibn Ahmad ibn Arfah al-Dasuqi, *Ḥāshiyat al-Dasūqī ʿalā al-Sharḥ al-Kabīr* (Cairo: Dar Ihya' al-Kutub al-ʿArabiyyah, n.d.), vol.1, p.509.
23. Salman ibn Khalf al-Baji, *Al-Muntaqā Sharḥ al-Muaṭṭā* (Beirut: Dar al-Kitab al-Islami, n.d.), vol.2, p.38.
24. Ibid.
25. Al-Bukhari, *Saḥīḥ*, vol.6, p.487.
26. Muslim, *Saḥīḥ*, vol.5, p.351.
27. Shihabuddin Ahmad ibn Ahmad al-Ramli, *Fatāwā* (Beirut: Dar al-Kutub al-ʿIlmiyyah, 1424 AH), vol.2, p.59.
28. Yahya ibn Sharaf al-Nawawi, *Al-Majmūʿah Sharḥ al-Muhadhdhab* (Cairo: Matbaʿah al-Muniriyyah, n.d.), vol.6, p.276.
29. Ibn Hajar, *Fatḥ al-Bārī*, vol.4, p.623.
30. Yusuf, *Caesarean Moon Births*, pp.47, 48.
31. Ibid., p.48.
32. Al-Shatibi, *Al-Muwāfaqāt*; that does not mean that al-Shatibi permitted use of calculations. He refuted them because of their imprecision and as requiring hardship for the unlettered Ummah.
33. Yusuf, *Caesarean Moon Births*, pp.28, 29.
34. Muslim, *Saḥīḥ*, vol.5, p.362.
35. Al-Bukhari, *Saḥīḥ*, vol.6, p.483.
36. Al-Tirmidhi, *Tafsīr al-Tirmidhī*, vol.3, p.114.
37. Abu Dawud, *Sunan*, vol.6, p.254.
38. Muslim, *Ṣaḥīḥ*, vol.5, p.345.
39. Abu ʿAbd al-Rahman Ahmad ibn ʿAli al-Nasā'ī, *Sunan al-Nasā'ī* (Beirut: Dar al-Kutub al-ʿIlmiyyah, 1991), vol.7, p.300.
40. Ibn Hajar, *Al-Talkhīṣ al-Ḥabīr fī Takhrīj Aḥādīth al-Rāfiʿī al-Kabīr*, 1st edn. (Beirut: Maktabah al-ʿIlmiyyah, 1989), vol.2, p.360.

41. Ibn Taymiyyah, *Fatāwā*, vol.6, p.590.
42. Ibid., vol.25, p.207.
43. Ibn Taymiyyah, *Al-Fatāwā al-Kubrā* (Beirut: Dar al-Kutub al-Ilmiyyah, n.d.), vol.1, p.62.
44. Ibid.
45. Ibid., vol.1, p.63.
46. Muhammad ibn Ahmad ibn Muhammad ʿAlish, *Manḥ al-Jalīl Sharḥ Mukhtaṣar Khalīl* (Beirut: Dar al-Fikr, n.d.), vol.2, pp.113-14.
47. Ibid., vol.2, p.114.
48. Al-Mubarak ibn Muhammad al-Jazri, *Al-Nihāyah fī Gharīb al-Ḥadīth wa al-Athar*, Tahir Ahmad al-Zawi and Mahmud Muhammad al-Tannakhi, eds. (Beirut: Maktabah al-Ilmiyyah, 1979), vol.2, p.205.
49. Ibn Taymiyyah, *Fatāwā*, vol.10, p.20.
50. Ibid.
51. Abu Dawud, *Sunan*, vol.10, p.412.
52. *Musnad Aḥmad*, vol.2, p.55.
523. Al-Bukhari, *Saḥīḥ*, vol.6, p.487.

CHAPTER 5

1. Al-Bukhari, *Saḥīḥ*, vol.6, p.481.
2. Muslim, *Saḥīḥ*, vol.5, p.355.
3. *Musnad Aḥmad*, vol.4, p.413.
4. Ibid., vol.5, p.251.
5. Ibid., vol.4, p.413.
6. Ibid., vol.19, p.137.
7. Ibid.
8. Ibid., vol.20, p.19.
9. Al-Bukhari, *Saḥīḥ*, vol.6, p.481.
10. *Musnad Aḥmad*, vol.20, p.48.
11. Ibid., vol.21, p.91.
12. Ibid., vol.33, p.28.
13. Ibid., vol.19, p.231.
14. Ibid., vol.38, p.355.
15. Ibid., vol.19, p.50.
16. Ibid., vol.41, p.396.
17. Al-Tirmidhi, *Sunan*, vol.3, p.106.
18. Ibid., vol.3, p.113.
19. Al-Darimi, *Sunan*, vol.5, p.166.
20. Abu Dawud, *Sunan*, vol.6, p.263.
21. Abu Hatim Muhammad ibn Habban, *Saḥīḥ Ibn Ḥabbān* (Beirut: Mu'assasat al-Risalah, 1993), vol.14, p.417.
22. Al-Bayhaqi, *Sunan al-Kubrā*, vol.4, p.207.

23. Ahmad Shafat, "A Study of *Aḥadīth* about the Determination of Islamic Dates," www.islamicperspectives.com, October 2003, p.13 .
24. Al-Nawawi, *Al-Majmūʿah Sharḥ al-Muhadhdhab*, vol.6, p.276.
25. Ibid.
26. Abu Dawud, *Sunan*, vol.6, p.256.
27. Al-Bayhaqi, *Al-Sunan al-Kubrā*, v.4, p.204.
28. Ibid.
29. Ibn Qudamah, *Al-Mughnī*, vol.3, p.7.
30. ʿAli ibn Ahamd ibn Hazm al-Anadulsi al-Zahiri, *Al-Maḥallā* (Beirut: Dar al-Fikr, n.d.), vol.4, p.445.
31. Ibn Qudamah, *Al-Mughnī*, vol.3, p.7.
32. Al-Nawawi, *Al-Majmūʿ*, vol.6, p.460.
33. Al-Azimabadi, *ʿAwn al-Maʿbūd*, vol.6, p.457.
34. Ibid.
35. Ibid.
36. Al-Bukhari, *Saḥīḥ*, vol.6, p.482.
37. Ibid., vol.6, p.483.
38. Al-Tirmidhi, *Saḥīḥ*, vol.3, p.116.
39. Shafat, *A Study of Aḥādīth*, p.16.
40. Ibid.
41. Salah Sultan, "Ru'yah ʿIlmīyah wa Tarbawīyah ḥawl Ru'yah al-Ahillah," presented to the Fiqh Council of North America, www.fiqhcouncil.org, 2006.
42. See Wahbah al-Zukhayli, *Al-Fiqh al-Islāmī wa Adillatuhu* (Beirut: Dar al-Fikr, 1997), p.1651.
43. Ibid., pp.1652-53.
44. Rida, *Tafsīr al-Manār*, vol.2, p.151.
45. An unpublished paper submitted to Shariʿah Scholars Association of North America (SSANA), Detroit meeting in 2000.
46. *The Soncino Talmud*, "Mas. Sanhedrin 10b," commentary on verse 2 (The Judaic Classical Library, Davka Corporation and Judaic Press, Inc. Electronic Edition 1991-1995).
47. Ibid., "Mas. Rosh HaShana," 20a, 1-4.
48. Ibid., 20b, 5-8.
49. Ibid.
50. www.JewishEncyclopedia.com/view.jsp?artid=43&letter=C&search=Calendar.
51. Ibid.
52. *Talmud*, "Mas. Sanhedrin," 12b, commentary to verse 33.
53. Ibid., "Mas. Chagigah," 14a, commentary to verse 43.
54. Ibid., "Mas. Sanhedrin," 11b, commentary to verse 9.
55. Ibid., "Mas. Sanhedrin," 11a, 22-29.
56. Ibid., "Mas. Sanhedrin," 11b, 7-10.
57. Ibid., "Mas. Arachin," 8b, commentary to verse 10.

58. www.JewishEncyclopedia.com/view.jsp?artid=43&letter=C&search=Calendar.
59. "The Jewish Calendar: A Closer Look," ibid.
60. Ibid.
61. See for details Muhammad Asad, *The Message of the Qur'an* (Lahore: Maktaba Jawhar ul-Uloom, n.d.), pp.264-65.
62. Al-Qaradawi, *Fatāwā Muʿāṣirah*, vol.2, pp.212-17.

CHAPTER 6

1. Al-Bukhari, *Saḥīḥ*, vol.6, p.478.
2. Muslim, *Saḥīḥ*, vol.5, p.342.
3. Ibid., vol.5, p.345.
4. Al-Nawawi, *Al-Majmūʿah*, vol.6, p.270.
5. *Al-Mawsūʿah al-Fiqhīyah,* vol.22, p.32.
6. Al-Nawawi, *Al-Majmūʿah*, vol.6, p.276.
7. Al-Baji, *Al-Muntaqā*, vol.2, p.38.
8. Ibn Daqiq, *Iḥkām*, vol.2, p.8.
9. *Al-Mawsūʿah al-Fiqhīyah*, vol.22, p.32.
10. Ibid., vol.22, p.33.
11. Abu Dawud, *Sunan,* vol.11, p.399.
12. *Musnad Aḥmad*, vol.36, p.27.
13. Abu Dawud, *Sunan,* vol.6, p.256.
14. Ibn Qudamah, *Al-Mughnī*, vol.3, p.7.
15. Al-Subki, *Fatāwā*, vol.1, p.209.
16. Ibid., vol.1, p.209.
17. Ibid.
18. Ibid., vol.1, pp.210-11.
19. Ibid., vol.1, p.210.
20. Ibid., vol.1, p.211.
21. Ibid., vol.1, p.217.
22. Al-Qardawi, *Fatāwā Muaʿsarah*, vol.2, p.222.
23. Zakariyya ibn Muhammad al-Ansari, *Al-Ghurar al-Baḥrīyah fī Sharḥ al-Bahjah al-Wardīyah* (Cairo: Maktabah al-Maimaniyyah, n.d.), vol.2, p.205.
24. Ibn Hajar, *Al-Talkhīs*, vol.2, p.360.
25. Ibn Daqiq, *Iḥkām*, vol.2, p.8.
26. Al-Hamwi, *Ghamz*, vol.2, p.65.
27. Ibn ʿAbidin, *Radd al-Mukhtār*, vol.2, p.387.
28. Ibn Hajar, *Fatḥ al-Bārī'*, vol.6, p.156.
29. Ahmad Shakir, *Awā'il al-Shuhūr al-ʿArabīyah* (Cairo: Maktabah Ibn Taymiyyah, 1987), see pages 7-17; the entire text is available at www.ahmad-muhammadshakir.blogspot.com, 2007.
30. Ibid.
31. Ibid.

32. Al-Zarqa, *Al-Fatāwā*, pp.163-64.
33. Al-Qardawi, *Fatāwā*, vol.2, pp.215-16.
34. Rida, *Tafsīr al-Manār*, vol. 2, p.151.
35. Sharaf al-Quda, a non-published paper on "Confirmation of the Lunar Month between the Prophetic Narrations and Modern Science," a paper in Arabic read at the SSANA meeting in Detroit, 2000.
36. Mustafa ʿAbd al-Samad Ahmad, *Taḥdīd Awā'il al-Shuhūr al-Qamarīyah* (Villanova, PA: Islamic Academy, n.d.), p.54 onward.
37. I am indebted to Dr. Jamal Badawi for this translation.

BIBLIOGRAPHY

Abou El Fadl, Khaled, "The Rules of Killing at War: An Inquiry into Classical Sources," *The Muslim World* 89 (1999).

———, "Speaking, Killing and Loving in God's Name," *The Hedgehog Review* (spring 2004), www.scholarofthehouse.org.

Abu Dawud, *Sunan Abū Dāwūd.* 4 vols. Beirut: Dar al-Fikr, n.d.

Abu Hanifah, Nuᶜman, *Al-Fiqh al-Akbar*, ed. Mulla ᶜAli al-Qari. Cairo: al-Babi, 1955. English trans. and comm. A. J. Wensick, *The Muslim Creed.* London: Frank Cass, 1965.

Abu Zahrah, Muhammad, *ᶜIlm Uṣūl al-Fiqh.* Cairo: Dar al-Fikr al-ᶜArabi, 1998.

al-Azimabadi, Muhammad Shams al-Haqq, *ᶜAwn al-Maᶜbud Sharḥ Abī Dāwūd.* Beirut: Dar al-Ilmiyyah, 1415 AH.

Afendi, ᶜAli Hayder Khawaja, *Durar al-Ḥikām Sharḥ Majallah al-Aḥkām.* trans. Fahmi al-Hussaini. Beirit: Dar al-Kutub alk-Ilmiyyah, n.d.

Ahmad, Mustafa ᶜAbd al-Samad, *Taḥdīd Awā'il al-Shuhūr al-Qamarīyah.* Villanova, PA: Islamic Academy, n.d.

al-Albani, M. Nasir al-Din, *Mukhtaṣar al-ᶜUlūw li al-ᶜAlīy al-Ghaffar*, supervised by Zahir al-Shawesh. Beirut, al-Maktab al-Islami, 2nd edn., 1412 AH.

ᶜAlish, Muhammad ibn Ahmad ibn Muhammad, *Manḥ al-Jalīl Sharḥ Mukhtaṣar Khalīl.* Beirut: Dar al-Fikr, n.d.

al-Alusi, Abu al-Thana' Shihab al-Din, *Rūḥ al-Maᶜānī fī Tafsīr al-Qur'ān al-ᶜAẓīm wa al-Sabᶜa al-Mathani.* 30 vols. Beirut: Dar Ihya' al-Turath al-ᶜArabi, n.d.

Anawati, Georges, "Attributes of God," *Encyclopedia of Religion.* ed. Lindsay Jones, 2nd edn. New York: MacMillian Reference Books: 2005.

al-Andalusi, Abu Hayyan Muhammad ibn Yusuf, *Tafsīr al-Baḥr al-Muḥīṭ.* Beirut: Dar al-Fikr, n.d.

Anderson, H., *Jesus and Christian Origins.* New York: Oxford University Press, 1964.

al-Ansari, Zakariyya ibn Muhammad, *Al-Ghurar al-Bahīyah fī Sharḥ al-Bahjah al-Wardīyah.* Cairo: Maktabah al-Maimaniyyah, n.d.

Arberry, A. J., *Revelation and Reason in Islam.* London: Allen & Unwin, 1965.

———, *Avicenna on Theology.* Westport, CT: Hyperion Press, 1979.

Armstrong, Karen, *A History of God.* New York: A. A. Knopf, 1994.

Asad, Muhammad, *The Message of the Qur'an.* Lahore: Maktaba Jawhar ul-Uloom, n.d.

al-Ashᶜari, Abu al-Hasan ᶜAli, *Kitāb al-Lumaᶜ*, ed. Father MacCarthy. Beirut: Imprimerie Catholique, 1953.

———, *Risālah fī Istiḥṣān al-Khawḍ fī al-Kalām*. Beirut: Imprimerie Catholique, 1953; briefly annotated English trans. Richard J. MacCarthy, *The Theology of al-Ashʿari: The Arabic Texts of al-Ashʿari's* Kitāb al-Lumaʿ *and* Riṣalāt Istiḥṣān al-Khawḍ fī ʿIlm al-Kalām. Beirut: Imprimerie Catholique, 1953.

———, *Al-Ibānah ʿan Uṣūl al-Diyānah*, ed. M. al-Khatib. Cairo: al-Matbaʿah al-Salafiyyah, n.d.; also ed. Fawqiyyah H. Mahmud. Cairo: Dar al-Ansar, 1977.

———, *Maqālāt al-Islāmīyīn wa Ikhtilāf al-Muṣallīn*, ed. M. ʿAbd al-Hamid. Beirut: al-Hikmah, 1994.

al-Awa, Adil, *Muntakhabāt Ismāʿīlīyah*. Damascus: Matbaʿah al-Jamiʿah al-Suriyya, 1957.

al-ʿAyni, Badr al-Din, *ʿUmdat al-Qārī*. Beirut: Dar al-Fikr, n.d.

al-Azhari, Abu Mansur Muhammad ibn Ahmad al-Harawi. *Tahdhīb al-Lughah*. Beirut: Dar Ihya' al-Turath al-ʿArabi, 2001.

al-Baji, Salman ibn Khalf, *Al-Muntaqā Sharḥ al-Muaṭṭa'*. Cairo: Dar al-Kitāb al-Islami, n.d.

al-Baladhiri, Ahmad ibn Yahya, *Ansāb al-Ashrāf*. Hamidullah edn. Cairo: Dar al-Maʿarif, n.d.

Baljon, J. M. S., "Qur'anic Anthropomorphism," in *Islamic Studies* 27. Islamabad: Islamic Research Institute (summer 1988).

al-Baqillani, Abu Bakar Muhammad ibn al-Tayyib, *al-Insaf fī ma Yajibu I'tiqaduh wa lā Yajuz al-Jahal bih*. Damascus: Izzat al-Husayni, 1950.

———, *Kitāb al-Tamhid*, ed. Yusuf al-Yasu'i. Beirut: al-Maktabah al-Sharqiyyah, 1957.

al-Bayhaqi, Abu Bakr Ahmad ibn al-Husayn ibn ʿAli, *Kitāb al-Asmā' wa al-Ṣifat*, ed. M. Z. al-Kawthari. Beirut: Dar al-Kutub al-ʿIlmiyyah, n.d.

Bearman, P. J., T. Bianquis, C. E. Bosworth, E. Van Donzel, and W. P. Heinrichs, *Encyclopedia of Islam*. Leiden: E.J. Brill, 1960–2005.

Bigg, Charles, "The Christian Platonists of Alexandria," *The 1886 Bampton Lectures*. Oxford: Clarendon Press, 1968.

Bin Bayyah, Abdullah ibn al-Shaykh al-Mahfudh, "Fatāwā," www.binbayyah.net, (2007).

Bright, John, *The Authority of the Old Testament*. New York: Abingdon Press, 1967.

Brown, Raymond E., "The Literal Sense of Scripture," in Raymond E. Brown, Joseph A. Fitzmyer, and Jerome Murphy eds., *The Jerome Biblical Commentary*. Englewood Cliffs, NJ: Prentice Hall, 1968.

Brunner, Emil, *The Christian Doctrine of God*, trans. Olive Wyon. Philadelphia: Westminster, 1974.

al-Bukhari, Abu ʿAbd Allah Muhammad, *Ṣaḥīḥ al-Bukhārī*, 1st edn. ed. Muhammad Zahir ibn Nasir al-Nasir. Beirut: Dar Tawq al-Najat, 1422 AH.

Chadwick, H., *Early Christian Thought and the Classical Tradition*. New York: Oxford University Press, 1966.

Danielou, J., *From Shadow to Reality*, trans. Dom W. Hibberd. Westminster, MD: Newman Press, 1960.

al-Darimi, Abu Muhammad ʿAbd Allah ibn ʿAbd al-Rahman, *Sunan al-Darimi*. Beirut: Dar al-Kitab al-ʿArabi, 1407 AH.

al-Dasuqi, Muhammad ibn Ahmad ibn Arfah, *Ḥāshiyat al-Dasūqī ʿalā al-Sharḥ al-Kabīr*. Cairo: Dar Ihya' al-Kutub al-ʿArabiyyah, n.d.

Denny, F., *An Introduction to Islam*, 2nd edn. New York: Macmillan, 1994.

Encarta, "Copernican System," http://encarta.msn.com/encyclopedia_761557138/copernican_system.html.

Encarta, "Solar System," http://encarta.msn.com/encyclopedia_761557663/solar_system.html.

Fakhry, Majid, *A History of Islamic Philosophy*. New York: Columbia University Press, 1970 .

al-Farabi, Abu Nasr, *Mabādi' Arā' Ahl al-Madīnah al-Faḍīlah*. Leiden: E. J. Brill, 1890.

———, *Al-Thamarāt al-Marḍīyah*. Leiden: E. J. Brill, 1895.

———, *Kitāb al-Millah wa Nuṣūṣ Ukhrā*, ed. Muhsin Mahdi. Beirut: Dar al-Mashriq, 1968.

al-Fayruzabadi, Muhammad ibn Yaʿqub, *Qāmūs al-Muḥīṭ*. Beirut: Mu'assasat al-Risalah, n.d.

al-Fayyumi, Muhammad, *Al-Miṣbāḥ al-Munīr fī Gharīb al-Sharḥ al-Kabīr*. Cairo: Matbaʿah al-ʿIlmiyyah, 1323 AH.

Ferguson, John, *Clement of Alexandria*. New York: Twayne Publishers, 1974.

Furnish, V. Paul, *Jesus According to Paul*. Cambridge, UK: Cambridge University Press, 1993.

Gibb, Hamilton A. R., *Mohammedanism: An Historical Survey*. London: Oxford University Press, 1972.

———, *Modern Trends in Islam*. New York: Octagon Books, 1972.

———, *Shorter Encyclopedia of Islam*. Leiden: Brill Academic Publishers, 1997.

Gibb, Hamilton A. R. et al. (eds.), *The Encyclopaedia of Islam*, new edn. Leiden: E. J. Brill, 1960.

Goldziher, Ignaz, *Introduction to Islamic Theology and Law*, trans. Andras and Ruth. Hamouri, with intro. & add. notes by Bernard Lewis. Princeton, NJ: Princeton University Press, 1981.

Grant, R. M., *The Letter and the Spirit*. London: SPCK, 1957.

Greer, Rowan A., trans., *Origen: The Classics of Western Spirituality Series*. New York: Paulist Press, 1979.

al-Hamwi, Ahmad ibn Muhammad, *Ghamz ʿAwn al-Basā'ir*. Beirut: Dar al-Kutub al-ʿIlmiyyah, n.d.

al-Hanbali, Abu al-Fath ʿAbd al-Hayy ibn al-ʿImad, *Shadharāt al-Dhahab fī Akhbār man Dhahab*. Cairo: 1350 AH.

Hick, John, *The Metaphor of God Incarnate*. London: SCM Press, 1993.

al-Husni, Abu Bakr Taqi al-Din, *Daf^c Shubah man Shabbaha wa Tamarrada wa Nasaba Dhālik ilā al-Imām Aḥmad.* Cairo: al-Halbi, 1350 AH.

Ibn ʿAbidin, Muhammad Amin ibn ʿUmar, *Radd al-Mukhtār ʿalā al-Dur al-Mukhtār.* Beirut: Dar al-Kutub al-ʿIlmiyyah, n.d.

Ibn Aqil, Abu al-Wafa ʿAli, *Kitāb al-Funun.* Beirut: Dar al-Mashriq, 1970–71.

Ibn al-ʿArabi, Abu Bakr Muhammad ibn ʿAbd Allah, *Aḥkām al-Qur'ān.* Beirut: Dar al-Kutub al-ʿIlmiyyah, n.d.).

Ibn ʿAtiyyah, Abu Muhammad ʿAbd al-Haqq ibn Ghalib, *Al-Muharrar al-Wajīz.* Beirut: Dar al-Kutub al-ʿIlmiyyah, 2002.

Ibn Faris, Abu al-Husayn Ahmad al-Qazwini al-Razi, *Maqāyīs al-Lughah,* ed. ʿAbd al-Salam Muhammad Harun. Damascus: Ittihad al-Kuttab al-ʿArab, 2000.

Ibn Habban, Abu Hatim Muhammad, *Saḥīḥ ibn Ḥabbān.* Beirut: Mu'assasat al-Risalah, 1993.

Ibn Hajar, Abu al-Fadal Shihab al-Din Ahmad ibn ʿAli, *Al-Talkhīṣ al-Ḥabīr fī Takhrīj Aḥādīth al-Rāfʿī al-Kabīr,* 1st edn. Beirut: Maktabah al-ʿIlmiyyah, 1989.

———, *Fatḥ al-Bārī fī Sharḥ al-Bukhārī.* Beirut: Dar al-Fikr, n.d.

Ibn Hanbal, Abu ʿAbd Allah Ahmad ibn Muhammad, *Musnad Aḥmad.* 6 vols. Cairo: Mu'assasat Qurtubah, n.d.

Ibn Hanbal, Ahmad, *Al-Radd ʿalā al-Zanādiqah wa al-Jahmīyah.* Cairo: Maktabah al-Salafiyyah, 1393 AH.

Ibn Hazm, Abu Muhammad ʿAli ibn Ahmad al-Andalusi al-Zahiri, *Al-Fasal fī al-Milal wa al-Niḥal.* Cairo: Maktabah al-Salam al-ʿAlmiyyah, n.d.

Ibn al-Jawzi, Abu al-Faraj ʿAbd al-Rahman ibn ʿAli, *Daf^c Shubah al-Tashbīh,* ed. al-Kawthari. Cairo: al-Maktabah al-Tawfiqiyyah, n.d.

Ibn Kathir, ʿImad al-Din Hafiz, *Al-Bidāyah wa al-Nihāyah.* Cairo: Matbaʿah al-Saʿadah, 1965.

———, *Tafsīr al-Qur'ān al-ʿAẓīm* (*Tafsīr Ibn Kathīr*). Beirut: Dar Ihya' al-Turath, n.d.

Ibn Manzur, Jamal al-Din, *Lisān al-ʿArab.* Beirut: Dar Ihya' al-Turath al-Islami, n.d.

Ibn al-Najjar, Taqi al-Din, *Sharḥ al-Kawkab al-Munīr.* Riyadh: Maktabat al-ʿAbaykan, 1997.

Ibn Nujaym, Zayn al-Din ibn Ibrahim, *Al-Baḥr al-Rā'iq Sharḥ Kanz al-Daqā'iq.* Beirut: Dar al-Kitāb al-Islami, n.d.

Ibn Qayyam al-Jawziyyah, Muhammad ibn Abi Bakr ibn Ayyub, *Iʿlām al-Mawaqqiʿīn ʿan Rabb al-ʿĀlamīn.* Beirut: Dar al-Jil, 1973.

———, *Ijtimāʿ al-Juyūsh al-Islāmīyah.* Beirut: Dar al-Kutub al-ʿIlmiyyah, 1984.

———, *Al-Sawā'iq al-Mursalah ʿalā al-Jahmīyah wa al-Muʿaṭṭalah.* Riyadh, Dar al-A'asimah, 1998.

Ibn Qudamah, ʿAbd Allah ibn Ahmad, *Al-Mughnī.* Beirut: Dar al-Fikr, 1405 AH.

Ibn Rajab al-Hanbali, Zayn al-Din ʿAbd al-Rahman ibn Ahmad. *Fatḥ al-Bārī.* Dammam: Dar Ibn al-Jawzi, 1422 AH.

Ibn Rushd, Muhammad ibn Ahmad ibn Muhammad, *Faṣl al-Maqāl fī mā bayna al-al-Ḥikmah wa al-Sharīʿah min al-Ittiṣāl*, ed. G. F. Hourani. Leiden: E. J. Brill, 1959.

Ibn Taymiyyah, Taqi al-Din Ahmad, *Naqd al-Mantaq*. Cairo: Matbaʿah al-Sunnah al-Muhammadiyyah, 1951.

———, *Taḥrīm al-Nazar fī Kutub Ahl al-Kalām*, trans. George Makdisi. London: Censure of Speculative Theology, 1962.

———, *Al-Rasā'il wa al-Masā'il*. Beirut: Dar al-Kutub al-ʿIlmiyyah, 1983.

———, *Fatāwā al-Kubrā*. Beirut: Dar al-Maʿrifah, 1386 AH.

———, *Kitāb al-Nubūwah*. Cairo: al-Maktabah al-Salafiyyah, 1386 AH.

———, *Minhaj al-Sunnah al-Nabawīyah*. Cairo: Mu'assasah Qurtubah, 1406 AH.

———, *Al-Risālah al-Tadummurīyah*. Beirut: al-Maktab al-Islami, n.d.

———, *Majmūʿ Fatāwā*, ed. ʿAbd al-Rahman ibn Muhammad ibn Qasim. Rabat: Maktabah al-Maʿarif, n.d.

al-ʿId, Ibn Daqiq, *Iḥkām al-Aḥkām Sharḥ ʿUmdat al-Aḥkām*. Cairo: Matbaʿah al-Sunnah al-Muhammadiyyah, n.d.

Ijaz, Abdur Rahman, "Axing of Methodology." www.hilalcommittee.com.

Jar Allah, Zahdi Hasan, *Al-Muʿtazilah*. Cairo: al-Mu'assasat al-ʿArabiyyah li al-Dirasat wa al-Nashr, 1947.

al-Jassas, Abu Bakr ibn ʿAli al-Razi, *Aḥkām al-Qur'ān*. Beirut: Dar al-Fikr, n.d.

al-Jawhari, Abu Nasr Ismaʿil ibn Hammad, *Al-Ṣiḥāḥ fī al-Lughah*. Beirut: Dar al-ʿIlm li al-Malayin, 1990.

al-Jawzi, Abu al-Faraj, *Tablīs Iblīs*. Beirut: Dar al-Kutub al-ʿIlmiyyah, 1993.

al-Jazri, al-Mubarak ibn Muhammad, *Al-Nihāyah fī Gharīb al-Athār*, eds. Tahir Ahmad al-Zawi and Mahmud Muhammad al-Tannakhi. Beirut: Maktabah al-ʿIlmiyyah, 1979.

al-Judayʿ, ʿAbd Allah ibn Yusuf, *Taysir ʿIlm Uṣūl al-Fiqh*. Beirut: Mu'assasat al-Rayyan, 2005.

Kawthari, M. Zahid, *Muqaddimah Tabyīn Kizb al-Muftara'*. Beirut: Dar al-Kitāb al-ʿArabi, 1984.

———, *Maqālāt al-Kawthari*. Cairo: Matbaʿah al-Anwār, n.d.

Kazi, A. K. and J. G. Flynn, trans., *Muslim Sects and Divisions: The Section on Muslim Sects in* Kitāb al-Milal wa al-Niḥal *by Muhammad ibn ʿAbd al-Karim Shahrastani*. London: Kegan Paul International, 1984.

al-Khalil, Ibn Ahmad, *Al-ʿAyn*. Beirut: Dar al-Kutub al-ʿIlmiyyah, 1996.

al-Kharshi, Muhammad ibn ʿAbd Allah, *Sharḥ Mukhtaṣar Khalīl li al-Kharshī*. Beirut: Dar al-Fikr, n.d.

al-Kirmani, Hamid al-Din, *Rāḥat al-ʿAql*. Cairo: Dar al-Fikr al-ʿArabi, 1952.

MacDonald, D. B., *Muslim Theology, Jurisprudence and Constitutional Theory*. Beirut: Khayats, 1965.

Madkur, Ibrahim, *Fī al-Falsafah al-Islāmīyah*. Cairo: Dar al-Maʿarif, 1976.

Maghraoui, Mokhtar, "An Islamic Legal Analysis of the Astronomical Determination of the Beginning of Ramadan." www.zawiyah.net and www.hilaldiscourse.net.

Mahmud, ʿAbd al-Halim, *Al-Tafkīr al-Falsafī fī al-Islām*. Cairo: Dar al-Maʿarif, 1984.

Al-Mawsūʿah al-Fiqhīyah. Kuwait: Ministry of Religious Affairs and Awqaf, 1967.

Mowlawi, Faysal, "Al-Sabab al-Sharʿī li Wujūb Ṣiyām Ramaḍān: Hal Huwa Dukhūl al-Shahr am Ru'yat al-Hilāl." Unfinished paper presented to the European Fiqh Council in May of 2007.

Muslim, Abu al-Hasan Muslim ibn al-Hajjaj, *Ṣaḥīḥ Muslim*. 5 vols. Beirut: Dar Ihya' al-Turath al-ʿArabi, n.d.

Nanji, Azim, "Ismaʿili Philosophy," in Seyyed Hossein Nasr and Oliver Leaman, eds., *History of Islamic Philosophy*, vol.1. New York: Routledge, 1996.

al-Nasafi, ʿAbd Allah ibn Ahmad, *Tafsīr al-Nasafī*. Beirut: Dar Ihya' al-Turath, n.d.

———, *Kanz al-Daqā'iq*. Beirut: Dar al-Nashr, n.d.

al-Nasafi, Abu al-Mu'in, *Baḥr al-Kalām*. Beirut: Dar al-Kutub al-ʿIlmiyyah, 2005.

al-Nasa'i, Abu ʿAbd al-Rahman Ahmad ibn ʿAli, *Sunan al-Nasa'i*. Beirut: Dar al-Kutub al-ʿIlmiyyah, 1991.

Nashshar, ʿAli Sami, *Nash'at al-Tafkīr al-Falsafī fī al-Islām*, 3rd edn. Cairo: Dar al-Maʿarif, 1965.

Nasr, Seyyed Hossein and Oliver Leaman, eds., *History of Islamic Philosophy*. New York: Routledge, 1996.

al-Nawawi, Yahya ibn Sharaf, *Al-Majmūʿah Sharḥ al-Muhadhdhab*. Cairo: Matbaʿah al-Muniriyyah, n.d.

Neil, Stephen, *The Interpretation of the New Testament 1861-1961*. New York: Oxford University Press, 1966.

Netton, R. I., *Allah Transcendent*. London & New York: Routledge, 1989.

Poonawala, Ismail K., "Ramadan," Microsoft Encarta. Reference Library 2005, Microsoft Corp., 1993–2004.

al-Qadi, ʿAbd al-Jabbar, *Al-Muḥīṭ bi al-Taklīf*. Cairo: al-Mu'assasat al-Masriyyah, 1965.

———, *Sharḥ Uṣūl al-Khamsah*, ed. ʿAbd al-Karim ʿUthman. Cairo: Maktabah Wahbah, 1965 .

al-Qalyubi, Ahmad Salamah and Ahmad al-Barlasi Umayrah, *Hashiyata Qalyūbī wa Umayrah*. Cairo: Dar Ihya' al-Kutub al-ʿArabiyyah, n.d.

al-Qaradawi, Yusuf, *Fatāwā Muʿāsirah*. Damascus: Dar al-Qalam, 1996.

Qari, Mulla ʿAli, *Mirqāt al-Mafātīḥ*. Cairo: Dar al-Fikr, n.d.

al-Qarrafi, Ahmad ibn Idris ibn ʿAbd al-Rahman Shihab al-Din, *Anwār al-Burūq fī Anwāʿ al-Furūq*. Beirut: Dar al-Kutub al-ʿIlmiyyah, 1998.

al-Qayrawani, Abu Muhammad ʿAbd Allah ibn Abi Zayd, *Al-Thamar al-Dāni', Risālah Abī Zayd al-Qayrawānī*. Beirut: Dar al-Fikr, n.d.

al-Quda, Sharaf, "Confirmation of the Lunar Month between the Prophetic Narrations and Modern Science." Unpublished paper in Arabic presented at the SSANA meeting in Detroit, Michigan, 1989.

al-Qurtubi, Abu ʿAbd Allah, *Tafsīr al-Qurṭubī.* Beirut: Dar al-Kutub al-ʿIlmiyyah, n.d.

Rahman, Fazlur, *Islam.* Chicago & London: University of Chicago Press, 1979.

———, "Sunni Islam," Microsoft Encarta. Reference Library 2005. Microsoft Corp., 1993–2004.

al-Ramli, Shihabuddin Ahmad ibn Ahmad, "Fatāwā." Beirut: Dar al-Kutub al-ʿIlmiyyah, 1424 AH.

al-Razi, Abu ʿAbd Allah Fakhr al-Din, *Asās al-Taqdīs.* Cairo: Matbaʿah Mustafa al-Babi, 1935.

———, *Iʿtiqadat Firaq al-Muslimīn wa al-Mushrikīn.* Cairo: Maktabah al-Nahdah, 1938.

———, *Al-Tafsīr al-Kabīr.* Beirut: Dar al-Fikr, 1978.

———, *Kitāb al-Arbaʿin fī Uṣūl al-Dīn.* Beirut: Dar al-Jil, 2004.

Rich, Tracey R., "The Jewish Calendar: A Closer Look." www.Jew-FAQ.org, (April 15, 2007).

Rida, Rashid, *Tafsīr al-Manār.* Beirut: Dar al-Kutub al-ʿIlmiyyah, 1999.

Rippin, Andrew, *Muslims: Their Religious Beliefs and Practices.* New York: Routledge, 1990.

Sachedina, Abdulaziz, "Warfare: The Use and Abuse of Jihad in the Muslim World." www.uga.edu/islam/jihad, 2006; see also www.islamiclearning.org, 2006.

Saint Augustine, *On Christian Doctrine*, trans. D. W. Robertson, Jr. New York: The Library of Liberal Arts, Bobbs-Merrill, 1958.

Sanders, E. P., *The Historical Figure of Jesus.* London: Allen Lane, 1993.

Seale, M. S., *Muslim Theology: A Study of Origins with Reference to the Church Fathers.* London: Lugac, 1964.

Sell, E. "God" (Muslim) in *Encyclopeadia of Religion and Ethics.* James Hastings, ed. New York: T. & T Clark Publishers, 1995.

Shafaat, Ahmad, "A Study of *Aḥādīth* about the Determination of Islamic Dates," www.islamic perspectives.com, (October 2003).

Shah, Zulfiqar Ali, "A Study of Anthropomorphism and Transcendence in the Bible and Qur'an: Scripture and God in the Judeo-Christian and Islamic Tradition." Unpublished Ph.D. thesis, University of Wales, Lampeter, 1997. www.hilal-discourse.net.

al-Shahrastani, ʿAbd al-Karim, *Al-Milal wa al-Niḥal.* London: Cureton, 1842.

Shakir, Ahmad, *Awā'il al-Shuhūr al-ʿArabīyah.* Cairo: Maktabah Ibn Taymiyyah, 1987.

Shatibi, Imam Abu Ishaq, *Al-Muwāfaqāt.* Beirut: Dar al-Kutub al-ʿIlmiyyah, n.d.

al-Shawkani, Muhammad ibn ʿAli, *Nayl al-Awṭār.* Beirut: Dar al-Jil, 1973.

———, *Al-Tuhaf fī Madhahib al-Salaf.* Cairo: Matbaʿah al-Imam, n.d.

Smith, W. B., *The Birth of the Gospel: A Study of the Origin and Purport of the Primitive Allegory of the Jews*. New York: Philosophical Library, 1957.
The Soncino Talmud, Judaic Classical Library. Davka Corp. & Judaic Press, www.judaicapress.com, (1991–1995).
al-Subki, Taqi al-Din, *Fatāwā al-Subkī*. Beirut: Dar al-Maᶜarif, n.d
al-Suyuti, Jalal al-Din, *Tafsīr al-Jalālayn*. Beirut: Dar Ihya' al-Turath al-ᶜArabi, n.d.
al-Tabari, Abu Jaᶜfar Muhammad ibn Jarir, *Jāmiᶜ al-Bayān fī Ta'wīl al-Qur'ān, Tafsīr al-Tabari*. Beirut: Mu'assasat al-Risalah, 2000.
al-Taftazani, Masᶜud ibn ᶜUmar, *Sharḥ al-Talwīḥ ᶜalā al-Tawḍīḥ*. Cairo: Maktabah Sabih, n.d.
Tamir, ᶜArif, *Khams Rasā'il Ismā'ilīyah*. Damascus: Dar al-Insaf, 1956.
———, *Ḥaqīqat Ikhwān al-Ṣafā' wa Khalān al-Wafā'*. Beirut: Catholic Press, 1957.
al-Tirmidhi, Abu ᶜIsa Muhammad, *Sunan al-Ṭirmidhī*. 5 vols. Beirut: Dar Ihya' al-Turath al-ᶜArabi, n.d.
Walzer, Richard, *Al-Farabi on the Perfect State*. Oxford: Clarendon, 1985.
Watt, W. M., "Early Discussions about the Qur'an," *Muslim World* 40 (1950).
———, *Islamic Philosophy and Theology*. Edinburgh: Edinburgh University Press, 1967.
———, *The Formative Period of Islamic Thought*. Edinburgh: Edinburgh University Press, 1973.
Wensick, A. J., *The Muslim Creed*. London: Frank Cass, 1965.
Wolfson, H. A., *The Philosophy of the Church Fathers*. 3rd rev. edn. Cambridge, MA: Harvard University Press, 1970.
Yusuf, Hamza, "Caesarean Moon Births." Berkeley: Zaytuna Institute, 2007.
Zaheer Uddin, "A Refutation to ISNA/Fiqh Council's Decision to Disregard the Qur'an and the Sunnah and to follow Astronomical Calculations for Beginning an Islamic Month." www.hilalsighting.org, issued Rajab 29, 1427 (August 24, 2006).
al-Zamakhshari, Abu al-Qasim Mahmud ibn ᶜUmar Jar Allah, *Asās al-Balāghah*. Beirut: Dar al-Fikr, n.d.
al-Zarqa', Mustafa, *Fatāwā Muṣṭafā al-Zarqā'*, ed. Majd Ahmad Makki, 2d edn. Damascus: Dar al-Qalam, 2001.
Zaydan, ᶜAbd al-Karim, *Al-Wajīz fī Uṣūl al-Fiqh*. Cairo: Dar al-Tawziᶜ wa al-Nashr al-Islamiyyah, 1993.
al-Zayla'i, Jamal al-Din ᶜAbd Allah ibn Yusuf, *Naṣb al-Rāyah fī Takhrīj Aḥādīth al-Hidāyah*. Cairo: Dar al-Hadith, n.d.
al-Zayla'i, ᶜUthman ibn ᶜAli, *Tabyīn al-Ḥaqā'iq Sharḥ Kanz al-Daqā'iq*. Beirut: Dar al-Kitab al-Islami, n.d.
al-Zukhayli, Wahbah, *Al-Fiqh al-Islāmī wa Adillatuhu*. Beirut: Dar al-Fikr, 1997.